INTRODUCTION
TO
FINANCIAL
ACCOUNTING

LEVIS D. McCULLERS
Associate Professor
Department of Accounting
University of Kentucky

RELMOND P VAN DANIKER
Associate Professor
Department of Accounting
University of Kentucky

MELVILLE PUBLISHING COMPANY

Los Angeles, California

INTRODUCTION TO FINANCIAL ACCOUNTING

Copyright © 1975, by John Wiley & Sons, Inc.
Published by *Melville Publishing Company*,
A Division of John Wiley & Sons, Inc.

Library of Congress Cataloging in Publication Data:

McCullers, Levis D 1936–
    Introduction to financial accounting.

    (Melville series on management, accounting, and
information systems)
    1. Accounting.   I. Van Daniker, Relmond P.,
1942–   joint author.   II. Title.

HF5635.M143   657   74–23201
ISBN 0–471–58365–0

Printed in the United States of America

10 9 8 7 6 5 4 3 2 1

To SHARON, THERESA, SUZANNE
and
DOLLY, JENNIFER, JANET
Who Shared

PREFACE

This textbook is a one-semester introduction to financial accounting and is designed to be used both by those majoring in accounting and by those who will take only one course in accounting. Consequently it will serve as an adequate preparation for those planning to take a course in managerial accounting.

The authors believe that most introductory accounting courses have been directed to accounting majors without sufficient consideration for nonaccounting majors. The result has often been a heavy concentration on the technical rather than on the conceptual aspects of accounting. The authors believe this approach has done little to facilitate the student's understanding of what accounting is and what accounting can do for him. This is especially important for the nonaccounting majors, who constitute the largest percentage of the enrollment in a beginning accounting course.

The authors have attempted to minimize the accounting techniques in order to present a blend of the conceptual and procedural aspects of accounting. In this manner, both the accounting major and nonmajor should acquire a better appreciation for the use of accounting for decision-making. In most chapters, technical details and procedural direction are provided because of our belief that the users of accounting information should have some awareness of how that information was developed. The emphasis, however, is on an understanding of the assumptions, principles, and alternative methods which influence the financial statements provided to external users. The objective is to explain accounting in a financial reporting framework rather than merely presenting financial ideas through the use of accounting terminology.

The text has been extensively classroom-tested both in regular day classes and in the evening class program. The instructors who have used the text in its preliminary form have ranged from senior professors to graduate students teaching their first course. The comments of the students and instructors contributed to the clarification of some illustrations and discussions.

The first two chapters cover the background of accounting, the necessity of accounting information for decision-making purposes, and the fundamental elements of financial statements. A set of financial statements is presented in Chapter 2 in order to illustrate the end product of the financial accounting process. The explanation of the statements is of a nontechnical nature such that the student is not overly confused. This approach provides the student with a frame of reference for the remainder of the text.

In keeping with our objective of minimizing the procedural details, we have presented the accounting process in two chapters. However, this does not mean that the procedural aspects have been eliminated to the point that accounting majors will suffer relative to their preparation for further accounting courses. This goal is partially achieved through the use of a comprehensive illustration of the accounting cycle in Chapter 4.

Chapters 5 through 11 present balance sheet items. In Chapter 5, cash is discussed primarily from an internal point of view. In addition, the chapter contains a comprehensive discussion of internal control. This was done because external statements are the result of internal activities and cash lends itself to a discussion of internal control better than other assets.

Chapters 6 through 9 present the other assets including material on present value concepts, revenue recognition, balance sheet valuation, and a brief discussion of consolidated financial statements. Chapter 10 is unique in that liabilities are presented through the use of the debt-equity classification criteria. Chapter 11 discusses owners' equity from the point of view of a proprietorship, a partnership, and a corporation.

Chapter 12 presents the Statement of Changes in Financial Position incorporating the recommendations of Accounting Principles Board Opinion Number 19. Chapter 13 is a further elaboration of the analysis of financial statements. Chapter 14 contains the background of the development of accounting principles and some contemporary issues. This chapter will serve to stimulate discussions of contemporary problems.

The last chapter is an overview of managerial accounting and may be eliminated in a one-semester financial course.

Some other features include:

1. Educational objectives at the beginning of each chapter. These objectives are statements of what the student should learn as a result of studying each chapter.

2. Discussion of accounting systems for each asset chapter. The systems' discussion serves to provide further insight relative to what is done internally in the process of developing external financial information.

3. Procedural and conceptual problems. The availability of both types of problems provides teacher flexibility. Those instructors who prefer an in-depth treatment of the accounting procedures can make more extensive use of the procedural problems while others may prefer to emphasize the conceptual problems. The number of procedural and conceptual problems is such that the instructor may emphasize procedures, concepts, or a blend of the two along the lines of the text.

4. Checklist of key figures included in the text after the table of contents.

In addition to the text, there is a *Student Study Guide*, prepared by Raymond J. Clay, which contains questions and short problems for each chapter. There is a set of solution paper, identified by problem number for the problems in each chapter of the text. There is also a solution and teachers' manual. Included in the solution manual is a series of examination questions for each chapter.

In addition to the foregoing material a readings book, *Contemporary Business Environment: Readings in Financial Accounting*, a selection of nontechnical articles organized to supplement the text, is available. There are articles related to each chapter to provide the student with additional insight into current issues in accounting.

ACKNOWLEDGMENTS

We are indebted to numerous reviewers whose comments and suggestions have contributed to the development of this text. We especially want to express our appreciation to John Buckley (University of California, Los Angeles), Alan Johnson (California State University, Hayward), Vernon Odmark (San Diego State University), and Charles Smith (Arizona State University). In addition, we are indebted to our colleagues at the University of Kentucky who assisted us in the class-testing of the book and made numerous helpful suggestions.

Most books require several drafts and this book was no exception. Throughout the development of the book we have been assisted by many typists; however, without the efforts of Alice Howell, Margot Dedas, and Shirley Springate this text could not have been completed.

Finally, we would like to thank the staff of Melville Publishing Company for their enthusiastic support and cooperation. These people include Bernard Scheier, Director of Production, and designer James F. Beggs. We cannot adequately express our appreciation to John Crain, Vice-President and Editor-in-Chief, for his support throughout the project; therefore we can only say thanks.

CONTENTS

1

IDENTIFICATION, MEASUREMENT, AND COMMUNICATION OF ECONOMIC INFORMATION

4

THE ADJUSTMENT AND CLOSING PROCESS   90

5

CASH   126

6

RECEIVABLES   152

7

INVENTORY   180

8

INVESTMENTS 206

9

FIXED ASSETS, INTANGIBLES, AND NATURAL RESOURCES 230

CONTENTS

14

SOURCES OF ACCOUNTING PRINCIPLES AND CONTEMPORARY ISSUES   372

15

MANAGERIAL ACCOUNTING: AN OVERVIEW   392

CHECKLIST
OF
KEY FIGURES

INTRODUCTION
TO
FINANCIAL
ACCOUNTING

1

IDENTIFICATION, MEASUREMENT, AND COMMUNICATION OF ECONOMIC INFORMATION

EDUCATIONAL OBJECTIVES

The material presented in this chapter is designed to achieve several educational objectives. These include:

1. An understanding of the purpose of accounting and the role of accounting information in the decision-making process.
2. A knowledge of the various users of accounting information and how the needs of those users differ.
3. An awareness of the criteria used for the identification of accounting information and the influence of the various accounting organizations on the development of those criteria.
4. An understanding of the different areas of accounting activities, such as internal, external, and governmental.

Decisions are an integral part of life. In order to make rational decisions, information is necessary. The more information that a decision-maker has, the greater the likelihood that the proper decision will be made; however, too much information can cause as many problems as too little information. Thus the decision-maker searches for the optimum amount of information for decision needs.

The information available to decision-makers is generally classified into two types:

1. Quantifiable. This type of information can be measured. Dollars, tons, pounds, number of soldiers available for a battle—these are measures of quantifiable information. Generally, the more measurable data available, the easier the decision. An individual can analyze the information (measurable data) and conclude that a particular course of action should be adopted.

2. Nonquantifiable. Much information cannot be measured, but nonetheless, it must be considered in reaching a decision. The reactions of people to particular courses of action cannot be quantified, but they must be considered in the decision-making process. The greater the quantity of nonquantifiable information, the more difficult the decision-making process.

Decision-Making Process

Decision-making, either by an individual or a business, involves five distinct phases:

1. Defining the problem. This step may be more difficult than expected. Few situations have neon signs indicating the problem.

2. Analyzing the problem. This step is completed in conjunction with step one. Analysis aids in clearly delineating the specific problem area.

3. Developing alternative solutions. As many alternative solutions as possible should be developed; however, consideration of every alternative in a particular situation is not possible for most people. Individuals are limited in their comprehension of the many variables involved in any decision. This phenomenon is known as the *Principle of Bounded Rationality*. The computer has greatly increased the number of variables that can be considered. In all probability, this has been the greatest contribution of computers to the decision-making process.

4. Deciding upon the best solution. From the many alternative solutions available, a choice must be made. Hopefully, the choice will be the best alternative. This probably will not be an optimal de-

cision. To have a true optimal, all possible contingencies must be evaluated, and problems are normally so complex that optimal decisions are impractical. Therefore, the choice usually is the best one given the variables considered. This practice of accepting less than optimal decisions is known as *satisficing*.

5. Converting the decision into effective action. Once the decision has been made, those actions necessary to achieve the desired objectives must be implemented.

The more quantitative information available to the decision-maker, the greater the number of alternatives that can be considered; however, most decisions involve, to some extent, the use of nonquantitative information. These feelings or beliefs must be considered and may generate alternative solutions that cannot be quantified. Thus the ultimate decision is based on a combination of quantifiable and nonquantifiable information.

As an example of the decision-making process, consider the following situation:

1. Defining the problem. You have no means of transportation and your job is 20 miles from your home.
2. Analyzing the problem. If you cannot get to work, you will lose your job. Since you have a family to support, you must have a job.
3. Considering the alternatives available:
   a. Ride in a car pool. There are no car pools in your area.
   b. Buy a motorcycle. The firm is a conservative firm and frowns on such modes of transportation.
   c. Buy a used car. Cost will be less than a new car, but the maintenance higher. Since you know little about car maintenance, a used car might present additional problems.
   d. Buy a new car. You are just starting your new job and you do not have enough money for a new car.
4. Deciding upon the best solution. Given the alternatives presented, you decide to buy a used car. You look for a used car which, at least on the surface, will not involve maintenance problems.
5. Converting the decision into effective action. You visit several used car dealers in order to look over the various models. Finally you decide on the car of your choice.

The preceding example illustrates the process that individuals should, and businessmen must, follow in making decisions. Individuals should do so because this is the way rational decisions are made. Since many decisions made by individuals are not subject to review, certain steps in the process, or the process itself, may be omitted. If the decision is a major one, affecting the individual over the long run, the likelihood is that this process will be followed. The good decision-maker is the individual who follows this process not only for major problems, but for minor problems as well.

Businessmen are in a different position from individuals. Their decisions are reviewed by others. In the case of corporations, the stockholders review the progress of management as evidenced by the decisions that have been made. If a decision is questioned, the executive must have reasons for the alternative selected. In some situations, the reviewers and the review process are not explicit. In every situation, however, the government requires records for the assessment of taxes. If a firm is in a regulated industry, e.g., public utilities, it must maintain records for the regulatory agency. Therefore, in any case, there are records which serve as the evidence of the outcome of the decision-making process.

Information Control

As stated previously, both quantitative and nonquantitative information must be available to make rational decisions. The larger the firm, the greater the need for control of this information. Too much information at the wrong place and the wrong time is as harmful as no information.

The need for information control in business has increased. At the beginning of this century, the nation was largely agricultural. The average size of enterprises was small and the entrepreneur, who operated the enterprise, knew all the essential facts about the business. This information was, most likely, carried in his head. Many financial records were informal since the owner was the only one who reviewed them.

In addition to the increased size of business, the passage of the Federal Income Tax Law of 1913 required the maintenance of additional records. Both affected information control profoundly. The volume of data needed to operate the business efficiently increased manyfold. As a general rule, the volume of information increases at a much faster rate than does the size of the business. The particular rate of expansion has been argued, but the fact that an information explosion exists is undeniable. Thus, the problem of receiving adequate information, which was hardly a problem for the small entrepreneur, has become a vital requirement for the managers of large organizations. This is true whether the organizations are profit-oriented, nonprofit, serving the needs of the community (hospitals, schools, etc.), or the entire nation (Department of Defense).

The types of information required by businessmen are the same as for individuals; however, businessmen must rely on the quantitative information to a greater extent than on information that cannot be measured. In order for the quantitative information to be useful, there must be some common form of communication. For a business this communication process is called *accounting*.

THE PURPOSE OF ACCOUNTING

The purpose of accounting is to provide quantitative information, primarily financial in nature, as an input to the decision-making process. In order to adequately serve this function, accounting involves the *processes* of (*a*) identifying, (*b*) measuring, and (*c*) communicating useful economic information to those who must make economic decisions.

The Users of Accounting Information

The processes of identifying, measuring, and communicating are the subject matter of this text. Therefore, prior to beginning the preliminary discussion of these topics, attention will be given to the users of accounting information. These users include: owners, creditors, managers, potential owners and creditors, employees, customers, governmental units, and the general public. Each of these user groups, and the types of decisions involved, will be briefly discussed.

Owners. The owners of a firm need accounting information to determine the success of the business. In the case of a corporation, the owners are interested in evaluating the effectiveness of management in the use and protection of the resources under its control. With this information, the owners can make a decision either to increase, decrease, or maintain their present investment in the firm.

Creditors. Accounting information is important to both prospective and present creditors. Their decisions concern the extension of credit as well as the terms of the credit arrangement. For example, some security may be required as protection before credit will be extended. They must also evaluate the likelihood that payment will be made when the debt comes due.

Managers. The management of a firm makes use of accounting information in assessing the results of operations, determining the need for cash, evaluating the potential for paying dividends, and making decisions relative to expanding or reducing the size and nature of the business.

Potential Owners. These users want to know how effectively management has exercised the stewardship function in the past. The potential owners are also concerned with the portion of ownership they may achieve and the expected return on their investment.

Employees. The employees use accounting information for wage expectations and negotiations. This can be carried out on an individual basis; however, the more probable situation is that the union representing the employees will use the information.

Customers.   The customers of the firm have a significant interest in whether the business can be expected to continue operations. An unfavorable financial situation may prompt customers to look elsewhere for a source of supply.

Governmental Units.   There are numerous governmental units, at the local, state, and federal level, which make use of accounting information for evaluating tax returns or assessing the taxes which are to be paid.

General Public.   This category includes all of the many other users of accounting information. In addition to the individual who may want such information about his personal affairs or for personal use, there are several groups who have an indirect interest because of their advisory or regulatory function. Among these are financial analysts, lawyers, financial news services, and governmental regulatory agencies, such as the Securities and Exchange Commission.

Obviously, the variety of users is a potential source of problems. Each of these user groups, as well as different users within a group, may have differing perceptions as to what information is important for its decision-making. There is a further problem in that all of the previously mentioned users, except management, are essentially external to the firm. Therefore, these users must rely upon someone else to provide the information needed since they do not have direct access to the records of the firm. Management, on the other hand, has ready access to, and control over, the accounting information which is generated for internal use. Also, the governmental units can require any information, within legal limits, which is necessary for their decisions.

The information needs of management and government have led to specialized areas of accounting activities. These areas are known as *internal*, *cost*, or *managerial* accounting, and *governmental* or *regulated industries* accounting. The accounting procedures involved in providing information employed in these areas are sufficiently different from the procedures involved in providing information to external users that they are presented in separate courses and texts. However, an overview of managerial accounting is presented in Chapter 15 of this text.

The purpose of this text is to explain those accounting activities which have particular relevance to external users. Accounting for external purposes is referred to as *financial* accounting. Some of the activities are carried out by the internal accountants because the external information is developed internally. The presentation is designed, however, for general purpose external users rather than for a specific internal use.

Accounting Processes

The accounting processes of identification, measurement, and communication involve economic transactions and events. In order for these processes to be operative, there must be some agreement as to the economic system and what constitutes transactions and events. This text discusses the accounting processes of those firms which are profit-oriented. In addition, resources are assumed to be limited; therefore, choices must be made. If the orientation of business changes, or if resources become unlimited, accounting activities will also change.

Identification of Economic Information. There are numerous bits of information which may be collected and used in the decision-making process. Accounting is only one source of information for decisions. The identification of information which is to be included in accounting relies upon the criteria of relevance and quantifiability.

Relevance. Since the volume of data which can be generated is infinite, the criterion of relevance is particularly important. The accountant is faced with the problem of selecting from the mass of data those things which are most likely to meet the needs of the different users of accounting information. These different users may have different goals; therefore, the accountant must know something about the decisions to be made on the basis of the information to be provided.

The number of users of accounting information, some of whom are not known at the time the data are identified, preclude the provision of special purpose information to each user. Rather, the accountant must analyze the needs of the various users and determine which information is common and relevant to all potential decision-makers.

The major problem in determining relevance is that different individuals have different views as to what is relevant. Consequently, questions may be raised concerning: relevant to whom; for what purpose; and during what time period? By making the data selection, the accountant assumes a significant role in the affairs of decision-makers. In fulfilling this role, he is guided by the ethical concepts of truth, justice, and fairness, which are considered to be fundamental to the accounting profession.

Quantifiability. There may be relevant data which cannot be quantified. Generally, such data do not become a part of the basic accounting information. As a general rule, accounting involves only those data which can be measured in dollars; however, dollars are not the only quantitative measure used by accountants. Certain other data may be presented as a supplement to the information which has been quantified in dollars.

Sometimes qualitative factors are assigned numbers in order to make certain evaluations. Such data may be provided by accountants, but they are not generally considered to be accounting information. Thus, when an accountant provides such information, he must be careful that the users understand the nature of the information.

Measurement of Economic Information. Once data have been identified as both relevant and quantifiable, the next step involves the actual measurement process. As stated earlier, measurement is usually in terms of dollars; thus the discussion in this section is not concerned with the unit of measure. Rather, its purpose is to describe the process of applying the measuring unit. The terms used to indicate the measurement standards include: objectivity or verifiability, and freedom from bias or neutrality. Since verifiability and neutrality are considered to be the more descriptive terms, they are discussed in this section.

Verifiability. The basic interpretation of this standard is that accounting information should be measured in such a manner that another accountant, with equal training and experience, can measure the same data and arrive at approximately the same results. The use of the term *approximately* is necessary because accounting is an art, not an exact science. As such, the identification and measurement processes are subject to judgment by the accountants involved. For example, two individuals may both determine precisely the cost of a machine, but reach different conclusions as to its expected useful life. Since both cost and the expected useful life of the machine are involved in the measurement of the success or failure of a business, the two individuals will arrive at different results.

Neutrality. The standard of neutrality is important to all users of accounting information, but especially to those users who are external to the firm. In the context of accounting, neutrality means that the information should not be designed for the benefit of a single user, or to give one user an advantage over other users. Within a firm, neutrality may be less important because, presumably, all of the users of accounting information have a common goal, i.e., the success of the business. However, neutrality is still important to management because biased information may result in erroneous decisions.

Communication of Economic Information. Once the subject matter of accounting has been identified and appropriately measured, the next process involves communicating this information to the decision-makers. The communication process is a specialized discipline which greatly exceeds the scope of this brief introduction. Therefore, the present purpose can best be served by noting that accounting information should be understandable, timely, uniform, and consistent.

Understandability. If accounting information is to serve a useful purpose, it must be understandable. The existence of many different users of the same information makes this task more difficult. For example, terminology which would be understandable for one category of users, such as bankers, may be misleading to others, such as a small scale investor. This does not imply, however, that accounting information must be presented so simply that its significance is lost. The users of the information have some obligation to be familiar with economic and accounting activities.

Timeliness. The more quickly accounting information is communicated to the users, the more likely it is to influence their decisions. The greater the delay in communicating, the greater the assurance that the information is correct. This implies that a trade-off may be necessary, i.e., some accuracy may be sacrificed in order to increase the timeliness of the information.

Accounting is primarily concerned with providing historical information. Nevertheless, that information should be made available as quickly as possible.

Uniformity. The information will be more understandable and useful if communicated in a relatively uniform manner. This pertains to the practices involved in obtaining the information as well as to the methods of reporting to the users.

Consistency. The standard of consistency implies that the information presented should be comparable with previous presentations. This does not mean that there can be no improvements in the communication process. Rather, this standard is based upon the view that the comparison of information of one time period with the information of another time period increases the value of the information. Obviously, if the two sets of information are not consistently prepared, comparability is impossible.

In summary, the components of identification, measurement, and communication have been discussed on an individual basis. From the conceptual standpoint, the various items — relevance, verifiability, etc. — are easy to present. However, the achievement of these objectives is difficult in practice. For example, the objectives of relevance and objectivity may conflict. Historical information is highly objective, but it may not be as relevant as current or forecasted information. There is a similar problem relative to neutrality and the user groups. Neutrality is difficult to achieve because decisions are made concerning the information to be reported, and the selection process itself may make the information non-neutral. Thus, users of accounting information must be aware of the impact that the selection process has on the information provided to them for decision-making purposes.

AREAS OF ACCOUNTING ACTIVITIES

There are three major categories of accounting activities in which the previously discussed criteria are utilized. These areas involve external, internal, and governmental reporting.

External Reporting

As previously mentioned, many users of financial information are external to the organization. They receive their information from management in the form of financial statements. Chapter 2 contains a discussion of the nature and format of these statements.

Financial statements are prepared by management for distribution to the interested users. Since the external users do not have access to the records of the firm, they have no direct means of ascertaining the validity of the statements. In order to provide the users with some assurance of validity, a common practice is to have the fairness of the statements certified by an external party. The certification is made by individuals known as *Certified Public Accountants* (CPA's), or external auditors. Note that the CPA only attests to the fairness of the statements. The purpose is not to ensure absolute accuracy. Judgments regarding fairness do not require that every item be reviewed. Furthermore, when the information was initially recorded by the company, certain assumptions and estimates were made which no amount of testing could prove to be completely accurate.

Unfortunately, many people believe that certified statements are accurate to the nth degree. For a CPA to determine absolute accuracy, every item would require review. Thus, the length of time required for verification would probably be equal to the period of time covered by the statements. By the time the information became available, the decision might have already been made; thus the information would be irrelevant. Consequently, some accuracy must be sacrificed for speed to provide relevant information.

Internal Reporting

In addition to the statements that are certified by the external auditor, other information is required by management. The format of the information for internal purposes varies with the objectives. Since management controls this type of information, the format for presentation is that which best serves the needs of management. Thus there are fewer specific guidelines for the presentation of accounting information for internal purposes than there are for information for external purposes.

The majority of internal accounting information is prepared by a *cost accountant.* This title is applied because most of the information involves the calculation of costs of particular operations. For example, before a firm decides to produce a product, it should calculate the cost of production to determine if the product can be sold at a profit. If the firm cannot sell the product at a profit then, probably, it should not be produced.

Other information for internal purposes involves the use of forecasted information. Management must have information concerning what is likely to happen in the next month, year, etc. This activity involves the preparation of budgets. The actual design of a budget, and the information contained in it, are a function of management's objective. This is the same type activity that individuals use when a budget for the next month is prepared in order to determine if all bills can be paid. Management follows the same practice. The only difference is that management prepares the budget formally, whereas the individual's preparation is usually informal.

Management would also like to be assured that appropriate information is received throughout the year. In order to achieve this objective, some accountants perform the same function for management that CPA's perform for external users. These accountants are called *internal auditors*, and they attempt to ensure accuracy on a daily basis. Thus, if the firm is large enough to require internal auditors, the job of the CPA is made considerably easier. The result is that management is satisfied that the internal information is accurate, and the CPA is more assured of the fairness of the statements for the external users.

Governmental Reporting

Financial information is also required by governmental agencies. The flexibility in the format of internal reports gives way to the formality of governmental reports. Formality is required so that compliance with the applicable law can be verified. For example, income tax returns are prepared on certain prescribed forms. These forms are designed so that all returns can be checked for mathematical accuracy and compliance with the income tax law. If everyone prepared his tax return in his own manner, chaos would result because no two returns would be similarly prepared.

Some firms offer services vital to the public interest, e.g., public utilities, airlines, etc. Because of the nature of their services, these firms must report to a regulatory agency, e.g., Federal Power Commission, Civil Aeronautics Board, etc. Reporting to governmental agencies, either federal or state, involves the same rigidity as reporting to the Internal Revenue Service.

The reports of regulated industries provide information for government statistics and serve as the basis for rate requests. In order to decide if an increase in rates is warranted, the regulatory commission must have comparable facts and figures. Since all firms in these industries use the same prescribed format, comparisons can be made.

The increased governmental regulations in business will undoubtedly increase the number of forms that must be prepared. This means increased job opportunities, either as government accountants or as accountants working for firms responsible for preparing these forms.

ORGANIZATIONS INFLUENCING ACCOUNTING THOUGHT

As previously described, accounting involves the processes of identifying, measuring, and communicating economic information. Since the entire accounting process is influenced by the business environment, the accountant must have some awareness of this in performing his function. Because of the complexity of the environment, accountants need certain guidelines for performing their function.

Several organizations which have a direct interest in the activity of accountants provide the guidelines. These groups provide information to accountants and users relative to those standards and practices which are regarded as generally acceptable and desirable. Among the more prominent organizations are the American Institute of Certified Public Accountants, Securities and Exchange Commission, American Accounting Association, National Association of Accountants, Financial Executives Institute, and the Internal Revenue Service.

American Institute of Certified Public Accountants (AICPA)

As the name implies, the AICPA is an organization composed of accountants that have qualified as CPA's. In order to obtain membership one must be a CPA, but it is not necessary for the individual to be engaged in the practice of public accountancy. Membership includes accountants in industry, government, and education.

Membership in the AICPA is purely voluntary and the AICPA has no legal authority. The individual states determine the qualifications necessary for licensing as a CPA and only the state can suspend the license; however, the AICPA can take certain administrative actions against any member whose conduct is considered detrimental to the profession. Even though individual states are responsible for licensing, all states administer the same uniform examination, prepared and graded by the AICPA.

The Securities and Exchange Commission (SEC)

The SEC, unlike the AICPA, has substantial legal authority. The Commission was created by Congress to administer the Securities Act of 1933 and the Securities Exchange Act of 1934. The SEC was given broad authority over financial information to be provided to the public by those companies which plan to raise capital by a public sale of securities. In addition, the Commission has jurisdiction over the annual reports required to be filed by those companies whose securities are listed for trading on stock exchanges.

American Accounting Association (AAA)

The AAA was begun in 1935, succeeding the American Association of University Instructors in Accounting which was established in 1916. This organization has a more academic orientation than the AICPA.

Financial Executives Institute (FEI)

The FEI has also made some contribution to the development of accounting thought. Membership in the FEI is primarily those individuals who occupy the highest financial positions within a large business. Accordingly, this organization has a major interest in financial reporting.

National Association of Accountants (NAA)

The NAA is composed primarily of industrial accountants. The contribution of this organization has been related to cost accounting and topics of general interest to management.

Internal Revenue Service (IRS)

The Internal Revenue Code has had only an indirect effect on accounting principles. The various revenue acts have no doubt contributed to the adoption of many accounting practices; however, the objective of the acts was not financial reporting, but rather providing revenue for the operation of government.

SUMMARY

The purpose of this chapter was to review the decision-making process and the relationship of accounting to the informational needs of that process. The decision-making process involves defining and analyzing the problem, accumulating information concerning alternatives, selecting the best alternative, and implementing the decision. Accounting aids in this process by providing quantitative information in the form of financial statements.

Accounting information is intended to be relevant to all the expected users, but not biased toward any one user. This information must be communicated in a uniform and consistent manner in order to be understandable. Since only one set of financial statements is provided for external users, there must be guidelines established for its preparation. These guidelines have been developed for the accounting profession by many groups interested in financial reporting. A more complete discussion of the influence of the AICPA, AAA, and SEC is presented in Chapter 14.

The remainder of this book, except for Chapter 15, is concerned with the guidelines for external, or financial, accounting. Since external users must rely principally upon published financial statements, Chapter 2 illustrates some typical financial statements. These statements are presented early in the text to provide the reader with a frame of reference for the remaining chapters.

QUESTIONS

1-1. (a) What are two types of information available to the decision-maker?
    (b) Which type makes decision-making easier? Why?

1-2. Name the five distinct phases of the decision-making process.

1-3. (a) What is the *Principle of Bounded Rationality*?
    (b) What has helped to overcome this phenomenon?

1-4. Why is it said that businessmen *must* follow the formal decision-making process?

1-5. Explain the correlation between the size of the firm and the volume of available information.

1-6. (a) What is the purpose of accounting?
    (b) What are the processes involved in accounting?

1-7. Name eight groups of users of accounting information.

1-8. Explain the criteria used in the identification of information which is to be included in accounting.

1-9. (a) What are four standards for measuring economic information?
    (b) Define verifiability and neutrality.

1-10. (a) What are four standards for communicating economic information?
    (b) Why is consistency important?

1-11. (a) Name the three major areas of accounting activities.
    (b) Distinguish between CPA's (external auditors) and internal auditors.

1-12. (a) Name six organizations which influence accounting thought.
    (b) Discuss the authority of the AICPA.

PROBLEMS

Conceptual Problems

1-1. *REQUIRED.* The chapter has a getting-to-work example which illustrates the steps involved in the decision-making process. From your personal experience, make up an example in which a decision is reached using the steps in the decision-making process.

1-2. *REQUIRED.* Match the group which uses accounting information with the types of decisions or information desired by placing the most appropriate letter in the blank.

(a) Owners   (e) Employees
(b) Creditors   (f) Customers
(c) Managers   (g) Governmental units
(d) Potential owners

—— (1) Evaluating tax returns or assessing taxes which are to be paid.
—— (2) Evaluating the effectiveness of management in the exercise of the stewardship function in the past.
—— (3) Wage expectations and negotiations.
—— (4) Evaluating the effectiveness of management in the use and protection of the resources under its control.
—— (5) Determining whether the business can be expected to continue operations.
—— (6) Need for information to carry out regulatory function.
—— (7) Determining the success of a business.
—— (8) Decisions relative to expanding or reducing the size of the business.
—— (9) Evaluating the likelihood that payment will be made when debt comes due.
——(10) Determining the need for cash.

1-3. Walter Conkern is an accountant for the Mason Corporation. He has to make a decision as to how inventories will be valued. Several methods are acceptable (to be discussed in a later chapter). Because of a friend who owns stock in the company, Walter chooses the method which results in the highest valuation. His belief is that this will result in an increase in the market price of the stock.

REQUIRED. Discuss this situation in terms of standards for the measurement of economic information.

1-4. The Mason Corporation (in Problem 1-3) selected Don Fritz as its auditor. When Don examined the inventories, he calculated a different value from the amount calculated by Walter.

REQUIRED. Discuss this situation in terms of standards for the measurement of economic information.

1-5. Herman Sermons, a CPA, was asked by the president of Smith Inc. to review the accounting system for cash receipts. Information available to him includes: sales made in the current month, collections from customers made in the current month, number of different customers, amounts owed by customers at the beginning of the current month, names of salesmen, and the amounts owed by customers at the end of the month. The president indicated that, in the past, the amount of cash received in any month was not known until the 15th day of the succeeding month. In addition, estimates of expected amounts for the past month ranged from $50,000 to $70,000.

REQUIRED. Discuss this situation in the context of the accounting processes of identification, measurement, and communication.

1-6. The president of a small manufacturing firm tells you that his company does not follow any set steps in arriving at a decision. He claims that decisions are joint efforts of several people in his firm. For instance, recently the company purchased a new machine used in the manufacturing process. The president instructed the purchasing department to buy the machine after studying the comparative costs of fixing the old machine, leasing, or buying a new one. These costs were furnished to him by the accounting department. The engineering department evaluated a report from the line foreman that the old machine was not functioning properly and then suggested that a cost analysis be completed.

REQUIRED. Refute the president's contention by naming the steps in his decision-making process and stating who performed each step.

1-7. Assume you are the vice-president in charge of plant operations for a medium-sized manufacturing firm. With respect to the following, what information would you expect the accounting department to supply?
   (a) Sales
   (b) Operating costs
   (c) Cash position
   (d) Inventories
   (e) Amounts owed by customers
   (f) Amounts owed to suppliers

1-8. You are the chief accountant for a laboratory which conducts scientific research for other large corporations. The company employs a number of highly educated scientists who seriously question the value of the information you generate. Explain the value of accounting information to a business enterprise.

1-9. An article appearing in the *Wall Street Journal* on May 24, 1973 included the following comment:
   The argument that a routine audit can't be expected to detect fraud is bolstered by the growing use of computers . . . Accounting sources say that the computer is causing auditors endless headaches and that only a few accounting firms are really coping with it.

REQUIRED. Express your views concerning possible ways in which auditors may begin to cope with the problems introduced by increased use of computer installations.

1-10. Comment briefly on each of the following statements. Your comments should indicate whether you agree or disagree with the statement as well as the reasons for your position:

(a) An accountant is merely another name for a bookkeeper. The only difference is that an accountant learns from formal education while a bookkeeper learns from practical experience.

(b) Accounting information is essentially historical in nature. Thus accounting data are useful in determining the past, but of little use in projections concerning the future.

(c) Computers will eventually replace the accountant.

(d) If it were not for the Internal Revenue Service, accounting reports would probably be issued no more frequently than every five years.

1-11. Included in the text is a section on the areas of accounting activities which discuss the concepts of internal and external reporting. From the discussion, it appears that a difference exists between the two types of reporting. When this fact is coupled with the knowledge that many users of accounting information have access only to external reports, a question arises. Are those who receive the information contained in internal reports receiving information which, if known by the external group, would cause them to change their view of the firm?

1-12. (a) A number of organizations influence accounting thought. This influence is in the form of professional pronouncements by accounting groups as well as government-originated documents. Of the six organizations listed in the text, which appears to possess the ability to be most influential? Which do you consider to be least influential?

(b) The accounting processes of identification, measurement, and communication are discussed in the text together with various criteria related to each. The eight criteria listed are as follows:

(1) Consistency   (5) Timeliness
(2) Neutrality   (6) Understandability
(3) Quantifiability   (7) Uniformity
(4) Relevance   (8) Verifiability

If you inherited $100,000 and decided to invest it in the stock of a publicly owned corporation, you would certainly want information concerning the operations of the company. Of the eight criteria listed here which three would you consider to be most important?

2

FINANCIAL STATEMENTS: AN OVERVIEW

EDUCATIONAL OBJECTIVES

The material presented in this chapter is designed to achieve several educational objectives. These include:

1. A basic knowledge of the format and content of financial statements, especially the balance sheet and income statement.

2. An understanding that the balance sheet is organized to draw attention to the relationship between current assets and current liabilities, total assets and total equities, and claims of creditors and claims of owners.

3. A knowledge of how the income statement is related to the balance sheet and how the amount of net income is the result of the matching of revenues and expenses.

4. Some basic insight into the methods for evaluating financial statements through the use of certain ratios.

Financial statements are the end product of the financial accounting process. These statements serve as the means of communicating information which has been identified and measured according to the criteria established in Chapter 1. The information should be communicated in an understandable, timely, uniform, and consistent manner. To accomplish these objectives, the statements should be prepared in accordance with generally accepted principles.

The term *generally accepted principles*, when applied to accounting, refers to the consensus, at a particular time, concerning the standards, methods, and practices to be employed in the identification, measurement, and communication of economic information. Therefore, these principles are referred to as *generally accepted accounting principles (GAAP)*.

Changes in the business environment affect the generally accepted accounting principles that are used. The principles discussed in this text are primarily related to profit-oriented firms; however, if the environment of business changes, i.e., from capitalism to socialism, the principles would change to reflect this difference. Therefore, accounting principles become generally accepted on the basis of experience, reason, and to a significant extent practical necessity.

Financial statements are presented in this chapter to afford the reader an opportunity to see the end product of the financial accounting process. There are two omissions in the statements included, (*a*) the Statement of Changes in Financial Position (discussed in Chapter 12), and (*b*) Footnotes to the Financial Statements (discussed in Chapter 13). The discussion will be of a nontechnical nature; however, the statements are similar to those published by any major corporation (see Appendix to Chapter 2). This discussion also provides an introduction to the terminology of accounting. The remaining chapters of the text will examine the technical considerations of the items presented.

BALANCE SHEET

A balance sheet (pp. 24–25) presents the financial position of a firm as of a particular date. Thus the statement is intended to provide a picture of a firm as though everything in the firm were momentarily at a standstill.

Note that on all balance sheets the total assets are equal to the total equities. This reflects the fundamental accounting equation of:

$$\text{Assets} = \text{Equities}$$

The asset section contains the goods and property the firm owns as well as claims which have not been collected. Thus assets are said to have future service potential for a firm.

Equities are the claims against the assets and indicate the source of the assets; e.g., owners invest funds and creditors loan money to the business. Because of the difference in the nature of these claims, the equity section is subdivided into the claims of creditors and claims of owners. Thus the fundamental equation could be expanded to:

$$\text{Assets} = \text{Liabilities} + \text{Capital}$$

The mechanics of the equation will be discussed and illustrated in Chapter 3.

The fact that a balance sheet was prepared implies that the financial position of Widget Company can be separated from all other firms. If, for some reason, the financial information of two firms could not be differentiated then a balance sheet could only be prepared for the two combined firms. A basic idea of accounting is that each business enterprise is an individual unit and is separate from those who supply the assets used. This is referred to as the *business entity assumption*.

Another point to be noted is the unit of measure. All items are stated in terms of dollars. There are no terms such as pounds or inches used as measures. All items are converted into dollars before they are presented on the balance sheet.

Working Capital

The term *working capital* refers to the excess of current assets over current liabilities. Current assets denote those assets which are reasonably expected to be converted into cash or consumed within the operating cycle or one year. The reason for distinguishing the current assets from all other assets is that current assets, in a cash form, will be used to pay the current obligations, or current liabilities.

Current liabilities are those obligations that must be paid, using current assets, within one year. They are distinguished from all other liabilities because of their relationship to current assets.

Current Assets. Each current asset of Widget Company will be briefly discussed in the order of appearance on the balance sheet. Current assets are normally listed in the order of *liquidity*, i. e., nearness to cash.

Cash. The title cash refers to *cash on hand* both at the company and in checking accounts. Note that the cash balance has increased from $100,000 to $145,000. The increase in cash may be an indication that the firm desires to maintain a more liquid position.

Marketable securities. These assets represent temporary investments of idle cash. These investments may be in such things as government securities which can be easily sold and, at the same time, earn interest. Since

The Widget Manufacturing Company, Inc.[a]
Balance Sheet
December 31, 1974
with Comparative Figures for 1973

ASSETS		1974		1973
Current Assets				
Cash		$ 145,000		$ 100,000
Marketable securities, at cost (market value: $60,000 in 1974 and $157,000 in 1973)		55,000		150,000
Receivables:				
Accounts receivable	$400,000		$425,000	
Less: Allowance for uncollectible accounts	20,000		25,000	
Net accounts receivable		380,000		400,000
Note receivable		10,000		8,000
Inventories (first-in, first-out basis)		462,000		454,000
Total Current Assets		$1,052,000		$1,112,000
Investments				
Land held for future plant site		$ –		$ 50,000
Investment in long-term securities (at cost)		100,000		100,000
Total Investments		$ 100,000		$ 150,000
Property Plant, and Equipment				
Land		$ 150,000		$ 100,000
Buildings	$825,000		$715,000	
Less: Accumulated depreciation	250,000	575,000	230,000	485,000
Machinery and equipment	$500,000		$450,000	
Less: Accumulated depreciation	90,000	410,000	75,000	375,000
Total Net Property, Plant, and Equipment		$1,135,000		$ 960,000

[a]The data included in the financial statements of the Widget Company are based upon an actual corporation. Rather than identify the company, the name Widget, which is a commonly used name in accounting literature, has been substituted.

ASSETS	1974	1973
ntangible and Other Assets		
Patents	$ 10,000	$ 10,000
Copyrights	5,000	6,000
Total Intangible and Other Assets	$ 15,000	$ 16,000
OTAL ASSETS	$2,302,000	$2,238,000

EQUITIES	1974	1973
Current Liabilities		
Accounts payable	$ 260,000	$ 210,000
Notes payable (due in 1 year)	10,000	40,000
Federal income tax payable	143,000	103,000
Sales commissions payable	28,000	26,000
Total Current Liabilities	$ 441,000	$ 379,000
ong-Term Liabilities		
Bonds payable	$ 100,000	$ 200,000
Total Liabilities	$ 541,000	$ 579,000
Owners' Equity		
Common stock, $10 par value Authorized shares 60,000; issued 50,000	$ 500,000	$ 500,000
Premium on common stock	150,000	150,000
Retained earnings	1,111,000	1,009,000
Total Owners' Equity	$1,761,000	$1,659,000
OTAL LIABILITIES AND OWNERS' EQUITY	$2,302,000	$2,238,000

these assets are relatively liquid, the market value is normally shown as a parenthetical comment.

Accounts receivable. This represents sales to customers for which payments have not been received. The total amount owed Widget Company on December 31, 1974 is $400,000; however, experience has shown that some customers fail to pay their bills. There are many reasons for this failure, such as financial difficulties or deliberate nonpayment. The potential value of the receivables is reduced to the expected collectible value by means of the allowance for uncollectible accounts. A failure to reduce the total receivables would result in an overstatement of their value. The accountant, in presenting information, tries to be conservative in his assumptions, and the overstatement of an asset is not a conservative position.

Notes receivable. This is the amount of receivables for which the company has received a promissory note. This is a written promise to pay a specified amount on some future day.

Inventories. This asset represents those items held for sale. For a manufacturing firm such as Widget Company, inventories may consist of raw materials, work in process, and finished goods. A retailer's inventory would normally be only finished goods.

Current Liabilities. Theoretically, current liabilities should be listed in the order of their maturity; however, this is difficult from a practical standpoint, because the items within an individual current liability have different maturity dates.

The balance sheet of Widget Company illustrates some of the common types of current liabilities. The accounts payable are the amounts owed to a firm's regular business creditors from whom merchandise was purchased on open account. This means that a promissory note was not given for the purchase. Any promissory notes that are given for purchases will appear under the caption, Notes Payable. Federal Income Tax Payable represents the amount due the government as taxes on the net income earned by the firm.

Sales Commissions Payable represents amounts owed the salesmen but unpaid as of December 31, 1974. The firm received the benefit of the sales in 1974 and any obligations, created by the sales, should be reflected on the financial statements in 1974.

Evaluation of the Current Position of Widget Company

The amount of working capital in 1974 was $611,000 as compared to $733,000 in 1973. This decrease of $122,000 resulted from a decrease in current assets of $60,000 and an increase in current liabilities of $62,000. Therefore, the ability of Widget Company to meet current obligations has been reduced, but it still appears to be sufficient.

The firm's ability to meet current obligations, as compared with other firms, can be evaluated by means of the *current ratio*. The current ratio is determined by dividing the current assets by the current liabilities. For Widget, the calculations are:

	1973	1974
$\dfrac{\text{Current assets}}{\text{Current liabilities}}$	$\dfrac{\$1,112,000}{\$379,000} = 2.93{:}1$	$\dfrac{\$1,052,000}{\$441,000} = 2.39{:}1$

In 1973, there were $2.93 of current assets for every dollar of current liabilities; however, that ratio decreased to $2.39 for every dollar of current liabilities in 1974. A decrease of this magnitude may indicate a lack of stability in the short run; however, a ratio of 2.39:1 is normally acceptable for a manufacturing firm.

The current ratio is an important consideration for the user interested in evaluating the short-term position of the firm. If the trend is downward, then a firm's liquidity position, or ability to pay current obligations, may be impaired. The reason for this is the time lag between the initial sale of the inventory and the ultimate collection of the receivable. There are no absolute standards for evaluating the current ratio. Various industries have different needs; consequently any analysis must take into consideration the nature of the business.

Noncurrent Assets

All assets, other than those classified as current, may be referred to as noncurrent assets. These assets are typically presented on the balance sheet under the captions: Investments, Property, Plant, and Equipment, and Intangibles. The amount shown on the balance sheet for these assets is normally the amount which was the cost at the time of purchase. In accounting terminology this is referred to as the use of *historical cost*.

Investments. These represent investments that are not expected to be converted into cash within one year. They are acquired for various reasons, e.g., land may be purchased for a future plant site. In 1973, Widget Company was holding land for a future plant site, and during 1974 this land was utilized for that purpose. Until the plant is built, the land is classified as an investment rather than property, plant, and equipment.

The investments in long-term securities are those held for more than one year. These securities differ from the marketable securities in that they are not *expected* to be sold in one year.

Property, Plant, and Equipment. These assets are frequently referred to as *Fixed Assets*. They represent assets used in the operation of the business and are not intended for sale.

Land. The land used as the site for a warehouse, plant, building, etc. is included in this category. As previously mentioned, any land not used for productive purposes is classified as an investment.

Buildings. The amount of the building is the historical cost, not the current market value. An argument against the use of market value is that the firm is a going concern; therefore, the assets were acquired for use in the business. The use of market value would indicate that the firm was either going out of business or planning to sell these assets. Perhaps, a more important argument against the use of market values is the difficulty of obtaining a verifiable market value for an asset in the absence of a sale.

As the asset is used, the historical cost is allocated to the periods benefitted by the asset. The total amount allocated is called *accumulated depreciation* and the historical cost is reduced on the balance sheet by the amount of the depreciation.

Machinery and equipment. These assets represent the machinery and equipment used in the productive process. They are classified separately from buildings in order to inform the readers as to the amount of machinery and equipment used. Since these assets are used in the production process, accumulated depreciation is also shown.

Intangible Assets. Intangibles may be defined as assets having no physical existence, but which provide potential value to the company. The Widget Company's intangibles, Patents and Copyrights represent two common intangible assets. Others could include franchises, goodwill, trademarks, and secret processes.

Patents may have potentially great value to the firm. The patent on the color television tube had substantial value for RCA in the early 1960s until the patent expired; however, for every valuable patent, there are numerous patents of insignificant value. Before presenting a patent on the balance sheet, the accountant must be reasonably certain that an asset exists. This is sometimes difficult in view of the lack of a physical existence. The amount which appears on the balance sheet is the original cost less the amount allocated to prior periods.

Copyrights represent expenditures made for the exclusive right to a literary work. As discussed in the section on patents, there must be reasonable certainty of future value before this asset would appear in the balance sheet.

In conclusion, the total assets increased from $2,238,000 to $2,302,000. The difference resulted from the change in many assets, and to consider only the total change in assets could be misleading to the user. Changes in individual asset categories is more informative than the change in total assets. The indication in this case is that Widget Company has reduced its current assets in favor of an expansion of its production facilities.

Equities

The preceding discussion was concerned with the various assets shown on the balance sheet. A common characteristic of assets is that they are things of value, which have the potential to provide benefit to the company in the future.

The portion of the balance sheet which reflects the sources of the assets is labeled *Equities*. There are two major categories, Liabilities and Owners' Equity, which enables users to distinguish the amount of assets provided by the creditors and the owners.

Liabilities. The major category Liabilities is a record of the amount of assets provided by creditors. The total amount is separated into two subtotals referred to as *current* and *long-term liabilities*. The purpose of the current category, as previously discussed, is to reflect more clearly those debt obligations which the company must pay within one year.

The long-term liabilities are those debts which will not require payment during the next year. Two of the common long-term liabilities are mortgages on property and bonds payable. Many companies obtain a substantial portion of their assets from the sale of bonds which may not be repaid for 20–25 years.

Owners' Equity. The category *Owners' Equity* is a record of the total amount of assets provided by the owners. This section of the balance sheet has several subsections depending upon the complexity of the firm. In the case of Widget Company, the owners purchased 50,000 shares of common stock with a par value of $10 for $650,000. (*Par value* is an arbitrary value assigned by the incorporators at the time the corporate charter is obtained.) Thus the owners were willing to pay $150,000 more than the par value, and this is shown as *Premium on Common Stock*. The *Retained Earnings* subsection refers to the amount of assets which were provided by profits earned by the firm. This item will be discussed more fully later in this chapter.

Both creditors and owners are interested in evaluating their relative position within the firm. The creditors expect the owners to provide security for the debt, and the owners want to know what use the company is making of borrowed funds. In order to evaluate the relative position of the creditors and owners, a ratio of liabilities to owners' equity is calculated. The debt-equity ratios for Widget Company as of December 31, 1973 and 1974 are:

$$
\frac{\text{Liabilities}}{\text{Owners' equity}}
$$

	1973	1974
$\dfrac{\text{Liabilities}}{\text{Owners' equity}}$	$\dfrac{\$579,000}{\$1,659,000} = 0.35{:}1$	$\dfrac{\$541,000}{\$1,761,000} = 0.31{:}1$

Thus, the owners of Widget Company as of December 31, 1974, have invested $1 for each $0.31 that the creditors have loaned the firm. This is a slight improvement for the creditors over the previous year when the owners had invested $1 for each $0.35 loaned the firm by the creditors.

Balance Sheet Valuation

There are several points to be considered relative to the dollar amounts shown on a balance sheet:

1. The dollar amount of the assets reflects different measurements. For example, cash is objectively determinable since the cash can be counted, whereas the net amount of the accounts receivable results from an estimate of the uncollectibles. In addition, fixed assets, such as plant and equipment, are shown at their historical cost less the portion of cost that has been allocated. Since there are a variety of methods for allocating the cost, the resulting numbers are influenced by the choice of methods.

2. The use of different measurements means that the total dollar amount of assets does not represent the current value of those assets. Rather the total asset amount is the sum of the various valuations of the assets.

3. The liabilities represent the amounts which will be paid to the creditors of the firm.

4. The amount of the owners' equity does not indicate the current value of the owners' claim. This is due to the fact that the owners' equity is a residual amount which results from the subtraction of the liabilities from the total assets. Since the assets do not represent current values, the owners' equity is not a current value.

Since the balance sheet does not reflect current values, some argue that it is not relevant for decision-making purposes. Others maintain that current values cannot be objectively determined and therefore should not be used. In addition, it may be argued that the dollar amounts shown on the balance sheet reflect in part decisions of management relative to the acquisition of assets, and thus current values are not appropriate. The purpose here is not to resolve this controversy but rather to point out that a user of the balance sheet must be aware of the measurement techniques employed in the balance sheet's preparation.

INCOME STATEMENT

As discussed in the preceding section, the balance sheet reflects the financial position of the company *at a specific time*. In contrast, the income

	1974		1973	
Net sales		$4,797,000		$4,400,000
Cost of goods sold		3,464,000		3,163,000
Gross profit on sales		$1,333,000		$1,237,000
Operating expenses:				
Selling expenses	$288,000		$275,000	
Administrative expenses	330,000		339,000	
General expenses	429,000	1,047,000	411,000	1,025,000
Operating income		$ 286,000		$ 212,000
Other income and expenses:				
Interest and dividends received		9,000		8,000
Income before income taxes		$ 295,000		$ 220,000
Federal and state income taxes		143,000		103,000
Net income		$ 152,000		$ 117,000
Earnings per share of common stock		$3.04		$2.34
Average number of shares of common stock outstanding		50,000		50,000

statement reflects the operations *for a period of time*. This is a clear indication of the different objectives of the two statements.

The purpose of the income statement is to provide information about the profitability or unprofitability of the business during a period. When statements for one or more previous periods are included, comparative evaluations can be made. The analyses of past performance provide insight into what may happen in the future.

The user of the income statement should be aware that the amount of reported net income is determined by applying certain generally accepted accounting principles. These principles pertain to such things as revenue recognition and matching.

Revenue recognition. Revenue is the inflow of assets to the firm resulting from the normal operation of the business. The normal accounting practice is to reflect the revenues (sales) at the time of sale, whether or not cash is received. The use of credit is common practice in business, and this results in a delay between the sale and the collection of cash.

When revenue is recognized at the point of sale rather than at the time of the collection of cash, this is referred to as the *accrual basis of accounting*. If revenue is recognized only upon the receipt of cash, this is called the *cash basis of accounting*. The generally accepted accounting practice is to use the accrual basis.

There are many problems in determining the appropriate revenue to be recognized in a given period of time. For example, when a land developer sells lots on an installment payment basis, should the amount of revenue be the total sales price or the amount collected in each period? Consequently, the user of an income statement should consider which revenue recognition procedures were employed by the firm.

Matching concept. Once the decision has been made regarding the revenue to be recognized, the next step requires that those expenses which were necessary to obtain the revenue be determined. The association of expenses with the revenue is referred to as *matching*. This means that expenses should be matched with the revenues which they helped generate.

The appropriate matching of revenues and expenses is important because the business is an ongoing concern but the results of operations are reported periodically. Therefore, if the results of the period are to be meaningful, revenues and expenses must be properly matched.

Thus net income is the result of particular decisions relative to revenue and expense recognition. The later chapters of this book will point out the various alternatives for expense recognition which can greatly influence the amount of reported net income.

In the following discussion of Widget Company, each major item in the income statement will be presented together with comments about the comparison between 1973 and 1974.

Net Sales. The caption *Net Sales* reflects the total amount of sales for the year, less any merchandise returned by customers and discounts allowed by the firm. Note that in the illustration, net sales increased by $397,000. This is a 9% increase in sales dollars. The increase may have resulted from two factors: (*a*) the volume of sales increased and/or (*b*) the selling price increased.

Cost of Goods Sold. The cost of the merchandise which was sold is deducted from net sales to determine *gross profit on sales*. The amount of gross profit indicates the average markup of selling price over cost. In 1973, Widget Company had a gross profit of 28% and in 1974, the ratio was just slightly below 28%. This may suggest that the company has a policy of selling merchandise at a 28% profit margin.

In the case of a retail store, the costs to be included in this category are the purchase price plus the delivery charges of the units sold. In a manufacturing company, the costs incurred in the production process are used. In a service business, e.g., doctors, lawyers, and accountants, there is no cost of goods sold since the generation of revenue does not involve the sale of merchandise.

Operating Expenses. The operating expenses of a business are all of the costs, except the cost of goods sold, which are a necessary part of the normal operations of the firm. These costs include selling, administrative, and general expenses. Some income statements provide only the total operating expense, while in others substantial information about the composition of the expenses is shown. The income statement of Widget Company has each of the major categories.

Selling expenses. This category includes all costs of selling the products, e.g., salaries and commissions paid to sales personnel, advertising, and entertainment of clients.

Administrative expenses. The expenses included in this category include the salaries of management, the salaries and other costs of the office staff, and other administrative costs of operating the business.

General expenses. As the title implies, this category includes everything which does not fit into one of the other categories. Some of the common costs included are rent, insurance, utilities, property taxes, etc.

Operating Income. The term *operating income* is used to designate the profit or loss resulting from the normal operations of the business. The calculation involves:

Net sales		$4,797,000
Less: Cost of goods sold	$3,464,000	
Operating expenses	1,047,000	4,511,000
Operating income		$ 286,000

This calculation is important because net operating income is the amount earned on the activities for which the firm was organized.

Other Income and Expense. There are situations in which a company may obtain revenue or incur expenses apart from its normal operations. The items under this caption are not necessarily unusual; rather they are not a part of the primary activities of the firm. Such amounts are shown separately in order that the results of normal operations may be compared from year to year. The inclusion of miscellaneous items would make comparisons difficult and might mislead the reader.

Income Taxes. Theoretically, the amount of taxes should be included under the caption *Operating Expenses.* Income taxes are omitted from that category because income must be determined prior to the calculation of taxes, because if there is no income, the firm does not incur any income tax expense.

Net Income. The net result of operating the firm can be determined only after all revenue and expense items have been considered. An excess of revenues over expenses is referred to as *net income.* If the expenses are larger than the revenues, the firm has a *net loss.* Note that the net income for Widget Company has increased from $117,000 in 1973 to $152,000 in 1974.

Evaluation of Net Income

Many readers of financial statements evaluate the rate of return management is earning on the assets used in the business. This information can be calculated by relating net income to total assets. The calculation is as follows:

$$\frac{\text{Rate of return}}{\text{on total assets}} = \frac{\text{Net income}}{\text{Total assets}} = \frac{\$152,000}{\$2,302,000} = 6.66\%$$

When this ratio is used for analytical purposes, the user should keep in mind the problems involved in income determination and asset valuation. Obviously, this *caveat* is applicable to any ratio generated from the financial statements.

In addition to the return on assets, the present owners, and potential investors as well, are interested in determining the rate of return on their investment in the business. The calculation of the rate of return on the owners' equity is:

$$\frac{\text{Rate of return}}{\text{on owners' equity}} = \frac{\text{Net income}}{\text{Owners' equity}} = \frac{\$152,000}{\$1,761,000} = 8.63\%$$

This return (8.63%) can then be compared by the owners to returns that could be earned from other types of investments. Consequently, if this return falls below that which can be obtained on other investments, the owners may decide to change their investment.

Earnings per Share of Common Stock. The calculation of earnings per share (EPS) of common stock is shown on the published income statements of publicly held corporations. This is a relatively new addition to the content of the income statement, and it emerged because investors considered this calculation to be extremely important. The earnings per share could be calculated by each user; but, because of certain complications which will be discussed in Chapter 13, misleading inferences might be drawn.

The simplified calculation of earnings per share which will be included at this point involves net income and the average number of outstanding shares of common stock:

$$\frac{\text{Earnings}}{\text{per share}} = \frac{\text{Net income}}{\text{Average number of shares of common stock}} = \frac{\$152,000}{50,000} = \$3.04$$

Thus the owners could receive $3.04 for every share of stock owned; however, the earnings per share figure does not indicate the amount each owner will actually receive. Distribution of profits to the owners of a corporation are called *dividends* and must be authorized by the board of directors of the corporation. Any portion of the earnings per share not paid to the owners increases the balance sheet category of *retained earnings*.

The amount of earnings per share has a major influence on the market price of the stock. While there are many other factors, such as the national attitude, the state of the economy, etc., which influence the stock market, the *price-earnings* ratio is perhaps the most common method of comparing the stock of one company to another. This is true because as the earnings per share increase, the market price of the stock should also increase.

A price-earnings ratio for Widget Company can be calculated by assuming that the market price of its stock is $50. The calculation is:

$$\frac{\text{Price-earnings}}{\text{ratio}} = \frac{\text{Market price}}{\text{Earnings per share}} = \frac{\$50}{\$3.04} = 16.5:1$$

This means that a prospective investor must be willing to pay 16.5 times the annual earnings per share to buy a share of Widget Company stock. Obviously the ratio will fluctuate based on the supply and demand for a share of Widget Company stock. Note that earnings per share is a statistic commonly used by investors; however, EPS should not be used as a substitute for a complete analysis of the income statement and balance sheet.

STATEMENT OF RETAINED EARNINGS

The purpose of the statement of retained earnings is to communicate information about changes in the earnings retained in the business. The

Retained Earnings Statement for Widget Company is provided for illustrative purposes.

The Widget Manufacturing Company, Inc.
Statement of Retained Earnings
for the Year 1974

Balance, January 1, 1974	$1,009,000
Add: Net income for the year	152,000
Total	$1,161,000
Less: Dividends paid	50,000
Balance, December 31, 1974	$1,111,000

The two major influences on retained earnings are net income and dividends. The addition of net income indicates that the owners' equity in the firm was increased by the total earnings for the period. The deduction of dividends reflects that, by action of the board of directors, a portion of the earnings was distributed, thereby reducing owners' equity. The ending balance of $1,111,000 is the *total amount* of earnings retained in the business since its beginning and is shown on the balance sheet presented earlier in the chapter.

SUMMARY

The purpose of this chapter was to present an overview of financial statements. Presentation was, to the extent possible, simplified and nontechnical. The chapter makes clear the need for an understanding of the accounting process by those who expect to use accounting information.

The remainder of the book is devoted to a more sophisticated presentation of accounting activities and practices. The following review will serve as a guide to a more meaningful understanding of the remaining chapters:

1. Financial statements are the end-product of the accounting process. This process is governed by those standards, practices, etc. which reflect the consensus of those who prepare and use the statements.
2. The financial accounting process and the statements in the United States use the dollar as the measurement unit. This unit *is presumed* to be stable, i.e., a dollar today is the same as a dollar in 1936.
3. Financial statements are historical documents and are not intended to include forecasts of what may happen.
4. Financial statements are designed for multiple user groups; therefore, the statements are intended to be neutral.
5. The information provided in the financial statements is summarized according to the expected needs of the users.

6. The total package of financial statements, i.e., balance sheet, income statement, statement of retained earnings, and the statement of changes in financial position (which will be discussed in Chapter 12) are related to each other. For example, assets are presented on the balance sheet because of their future service potential. When there is some expiration of the potential, the cost of the expired portion becomes a deduction on the income statement.

10 YEAR FINANCIAL SUMMARY
Levi Strauss & Co. and Subsidiaries
(Dollar Amounts in Millions Except Per Share Data)

	1973	1972	1971
Net Sales	$653.0	$504.1	$432.0
Income Before Taxes (1)	33.8	48.1	35.7
Net Income	11.9	25.0	19.7
Earnings Retained in the Business	6.6	20.9	16.3
Cash Flow Retained in the Business (2)	16.5	28.0	21.9
Income Before Tax as % of Sales	5.2%	9.5%	8.3%
Net Income as % of Sales	1.8%	5.0%	4.6%
Net Income as % of Beginning Stockholders' Equity	7.0%	16.8%	23.2%
Current Assets	$305.5	$252.4	$202.8
Current Liabilities	155.7	98.2	67.9
Working Capital	149.8	154.2	134.9
Ratio of Current Assets to Current Liabilities	1.96/1	2.57/1	2.99/1
Long-Term Debt (3)	$ 48.1	$ 37.6	$ 28.4
Stockholders' Equity	176.4	169.7	148.8
Capital Expenditures	$ 28.8	$ 17.6	$ 15.6
Depreciation & Amortization	9.8	7.1	5.6
Property, Plant & Equipment—Net	68.0	48.0	39.6
Number of Employees	29,141	25,137	21,383

PER SHARE DATA:			
Net Income (4)	$ 1.09	$ 2.30	$ 1.86
Cash Dividends Declared	.48	.38	.32
Book Value (on Shares Outstanding at Year End)	16.21	15.60	13.68
Market Price Range	49 3/4-16 5/8	59 3/4-40 3/8	64 1/4-33 1/4
Shares Outstanding (4)	10,880,080	10,880,080	10,586,000

(1) *Net of minority interest in net income of consolidated subsidiaries.*
(2) *Net income plus depreciation and amortization minus dividends declared.*
(3) *Excludes short term portion.*
(4) *Based on weighted average of shares outstanding during the year.*

1970	1969	1968	1967	1966	1965	1964
$349.5	$269.0	$ 213.5	$176.5	$164.3	$136.5	$111.5
37.7	31.7	26.6	14.8	15.5	15.3	15.8
18.6	14.7	12.1	7.8	7.9	7.6	6.9
15.7	12.0	10.4	5.8	5.8	5.1	5.0
20.1	14.9	12.8	8.0	7.4	6.2	5.6
10.8%	11.8%	12.5%	8.4%	9.4%	11.2%	14.2%
5.3%	5.5%	5.7%	4.4%	4.8%	5.6%	6.2%
26.8%	25.6%	25.8%	19.0%	23.2%	26.4%	29.1%
$169.0	$131.0	$ 94.9	$ 78.0	$ 71.4	$ 48.7	$ 39.7
87.9	57.5	29.6	23.3	21.3	20.3	14.4
81.1	73.5	65.3	54.7	50.1	28.4	25.3
1.92/1	2.28/1	3.21/1	3.35/1	3.35/1	2.40/1	2.76/1
$ 25.4	$ 22.6	$ 22.2	$ 22.1	$ 22.2	$ 1.8	$ 1.9
85.0	69.3	57.4	46.9	.41.0	34.0	28.8
$ 14.5	$ 6.9	$ 2.4	$ 3.9	$ 4.5	$ 2.4	$ 2.9
4.4	2.9	2.4	2.2	1.6	1.1	.6
29.2	19.4	15.4	14.6	13.5	8.8	7.3
18,900	16,466	14,067	12,527	13,738	9,340	7,333
$ 1.92	$ 1.50	$ 1.23	$ .79	$ .81	$ .78	$ .71
.30	.28	.18	.21	.21	.20	.20
8.66	7.07	5.85	4.79	4.20	3.49	2.96
—	—	—	—	—	—	—
9,661,000	9,754,000	9,809,000	9,791,000	9,771,000	9,756,000	9,711,000

CONSOLIDATED STATEMENT OF INCOME
Levi Strauss & Co. and Subsidiaries

	Year (52 Weeks) Ended	
	November 25, 1973	November 26, 1972
Net sales	$653,042,000	$504,104,000
Cost of goods sold	468,650,000	343,829,000
Gross profit	$184,392,000	$160,275,000
Marketing, general and administrative expenses	139,179,000	107,068,000
Operating income	$ 45,213,000	$ 53,207,000
Interest expense	10,133,000	4,328,000
Other expense (net of other income)	1,272,000	810,000
Income before taxes	$ 33,808,000	$ 48,069,000
Provision for taxes on income	21,952,000	23,046,000
Net income	$ 11,856,000	$ 25,023,000
Net income per share	$ 1.09	$ 2.30
Average number of shares of common stock outstanding	10,880,080	10,880,080

CONSOLIDATED STATEMENT OF STOCKHOLDERS' EQUITY
Levi Strauss & Co. and Subsidiaries

	COMMON STOCK		Additional	
	Shares	Stated Value	Paid-In Capital	Retained Earnings
Balance, November 28, 1971	10,880,080	$48,960,000	$43,563,000	$56,307,000
Net income				25,023,000
Cash dividends declared ($0.38 per share)				(4,129,000)
Balance, November 26, 1972	10,880,080	$48,960,000	$43,563,000	$77,201,000
Net income				11,856,000
Cash dividends declared ($0.48 per share)				(5,222,000)
Balance, November 25, 1973	10,880,080	$48,960,000	$43,563,000	$83,835,000

The accompanying accounting policies and notes to consolidated financial statements are an integral part of these statements.

FINANCIAL STATEMENTS: AN OVERVIEW

CONSOLIDATED STATEMENT OF
CHANGES IN FINANCIAL POSITION

Levi Strauss & Co. and Subsidiaries

	Year (52 Weeks) Ended	
	November 25, 1973	November 26, 1972
WORKING CAPITAL PROVIDED BY:		
Operations:		
Net income	$11,856,000	$25,023,000
Add items not currently involving working capital:		
Depreciation and amortization	9,836,000	7,144,000
Other items, net	1,251,000	588,000
Working capital provided by operations	$22,943,000	$32,755,000
Proceeds from sale of home office facilities less expenses	—	3,276,000
Proceeds from long-term debt	24,984,000	12,591,000
Working capital provided	$47,927,000	$48,622,000
WORKING CAPITAL USED FOR:		
Additions to property, plant and equipment	$28,812,0C	$17,554,000
Cash dividends declared	5,222,000	4,129,000
Reductions in long-term debt	14,617,000	3,372,000
Excess of purchase price over net assets of subsidiaries		
acquired and purchase of minority interest	2,353,000	3,524,000
Other transactions, net	1,182,000	766,000
Working capital used	$52,186,000	$29,345,000
Increase (decrease) in working capital	($ 4,259,000)	$19,277,000
INCREASE (DECREASE) IN WORKING CAPITAL:		
Cash and temporary cash investments	$ 968,000	($ 427,000)
Trade receivables, less allowances	27,960,000	9,317,000
Inventories	22,553,000	36,491,000
Other current assets	1,677,000	4,253,000
Current maturities of long-term debt		
and due to banks	(21,297,000)	(22,753,000)
Accounts payable and accrued liabilities	(36,216,000)	1,397,000
Other current liabilities	96,000	(9,001,000)
	($ 4,259,000)	$19,277,000

The accompanying accounting policies and notes to consolidated financial statements are an integral part of this statement.

CONSOLIDATED BALANCE SHEETS
Levi Strauss & Co. and Subsidiaries

ASSETS	November 25, 1973	November 26, 1972
Current Assets:		
Cash and temporary cash investments	$ 15,940,000	$ 14,972,000
Trade receivables (less allowances for doubtful accounts — 1973 — $3,335,000; 1972 — $1,115,000)	101,764,000	73,804,000
Inventories	177,038,000	154,485,000
Other current assets	10,801,000	9,124,000
Total current assets	$305,543,000	$252,385,000
Property, Plant and Equipment — Net	68,010,000	47,998,000
Other Assets	9,106,000	6,678,000
	$382,659,000	$307,061,000
LIABILITIES AND STOCKHOLDERS' EQUITY		
Current Liabilities:		
Current maturities of long-term debt	$ 3,458,000	$ 799,000
Due to banks under notes payable and overdraft agreements	53,688,000	35,050,000
Accounts payable and accrued liabilities	75,833,000	39,617,000
Salaries, wages and employee benefits	12,569,000	9,653,000
Taxes based on income	8,809,000	12,039,000
Dividends payable	1,306,000	1,088,000
Total current liabilities	$155,663,000	$ 98,246,000
Long-Term Debt — Less current maturities	$ 48,110,000	$ 37,604,000
Deferred Items	$ 2,528,000	$ 1,487,000
Stockholders' Equity:		
Common stock — $1.00 par value: 20,000,000 shares authorized: 10,880,080 shares outstanding (at stated value)	$ 48,960,000	$ 48,960,000
Additional paid-in capital	43,563,000	43,563,000
Retained earnings	83,835,000	77,201,000
Total stockholders' equity	$176,358,000	$169,724,000
	$382,659,000	$307,061,000

The accompanying accounting policies and notes to consolidated financial statements are an integral part of these balance sheets.

FINANCIAL STATEMENTS: AN OVERVIEW

ACCOUNTING POLICIES

PRINCIPLES OF CONSOLIDATION

The consolidated financial statements include the accounts of the Company and all subsidiaries. Intercompany accounts and transactions have been eliminated in consolidation.

TRANSLATION OF FOREIGN CURRENCIES

The Company's operations outside the United States accounted for 39% and 32% of consolidated assets and sales in 1973 and 39% and 29% in 1972. The accounts of such operations, which are maintained in other currencies, have been translated into U.S. dollars as follows:

Monetary assets and liabilities (including long-term debt) have been translated at current exchange rates at the respective year-ends.

Non-monetary items, principally inventories and property and equipment (and related reserves) have been translated at the exchange rates in effect at the time such assets were acquired. Stockholders' Equity, except for the current years' results, has been translated at appropriate historical exchange rates.

Income and expenses, other than depreciation and cost of goods sold, which have been translated at the historical exchange rates that apply to the related assets, have been translated at a weighted average exchange rate for the year. The resulting exchange adjustments, including gains and losses on foreign exchange contracts (all of which have been realized) were not material in 1973 and 1972.

INVENTORY VALUATION

Inventories are valued substantially at the lower of average cost or market.

DEPRECIATION METHODS

For financial reporting purposes, accelerated depreciation methods are applied on approximately 56% of depreciable assets and the straight line method is used on the remaining depreciable assets. Rates are based on the estimated useful lives of the property.

AMORTIZATION OF INTANGIBLES

Intangibles, including trademarks, trade names, goodwill and covenants-not-to-compete are being amortized on a straight line basis over periods of expected benefit which range principally from three to five years.

RESEARCH AND DEVELOPMENT

Research and development expenditures are charged to current operations as incurred.

RETIREMENT PLANS

The pension costs for the year include the cost of current service and the amortization of past service costs over ten years. Pension costs are funded as accrued.

INCOME TAXES

Deferred income taxes are provided for all significant timing differences between financial and tax reporting. The Company does not provide for taxes which would be payable if the net cumulative undistributed earnings at November 25, 1973 of its foreign subsidiaries ($14,400,000) and of its Domestic International Sales Corporation (DISC) ($1,451,000) were remitted to the Company. The Company intends to permanently re-invest such earnings either in the operations of the subsidiary companies or, in the case of the DISC, in accordance with existing tax regulations and thus defer indefinitely the tax liability.

The Company reduces the provision for federal income taxes currently for the investment tax credit on qualified property additions.

NET INCOME PER SHARE

Earnings per share are calculated on the basis of the weighted average number of common shares outstanding for the period. Exercise of stock options granted would not have a dilutive effect on net income per share.

NOTES TO CONSOLIDATED FINANCIAL STATEMENTS

ACQUISITIONS

In 1972, the Company exchanged 66,000 shares of its common stock for all the outstanding stock of Miller Belts Ltd., Inc. This transaction was accounted for as a pooling of interests. Also during 1972, the Company purchased an additional 20% of the outstanding shares of its Canadian subsidiary, GWG Limited, for $2,702,000 thereby increasing its ownership to approximately 99%.

During 1973, the Company acquired the operations of its Portuguese distributor, a Canadian belt manufacturer and three domestic garment manufacturers for a total purchase price of $4,860,000.

The goodwill arising from 1972 and 1973 purchases is being amortized over five years.

INVENTORIES

Inventories consist of:

	1973	1972
Finished goods	$ 96,166,000	$ 92,606,000
Raw materials and work-in-process	80,872,000	61,879,000
	$177,038,000	$154,485,000

PROPERTY, PLANT AND EQUIPMENT

Property, plant and equipment stated at cost less accumulated depreciation consist of:

	1973	1972
Land	$ 3,123,000	$ 2,958,000
Buildings and leasehold improvements	34,443,000	27,334,000
Machinery and equipment	52,842,000	42,283,000
Construction in progress	12,566,000	2,713,000
Total cost	$102,974,000	$ 75,288,000
Less accumulated depreciation	34,964,000	27,290,000
	$ 68,010,000	$ 47,998,000

Depreciation charges for fiscal years 1973 and 1972 amounted to $8,302,000 and $6,393,000, respectively. Property, plant and equipment includes $9,600,000 and $3,500,000 for 1973 and 1972 resulting from the capitalization of leases, less accumulated depreciation.

LONG-TERM DEBT

Long-term debt is summarized below:

	1973	1972
Secured by properties:		
Notes payable, 4% to 9-3/4%, due in installments through 1990	$ 8,361,000	$ 3,952,000
Capitalized leases due through 1993 of $14,880,000, less unexpended construction funds of $5,079,000 held by trustees	9,801,000	3,960,000
Other indebtedness, 6% to 13%	692,000	1,198,000
	$18,854,000	$ 9,110,000
Unsecured:		
Notes payable to an insurance company, 7-5/8%, due in annual installments of $1,550,000 commencing 1974	27,800,000	17,800,000
Eurodollar loan, 7-1/8%	—	10,000,000
Other indebtedness, 3% to 12-1/4%	4,914,000	1,493,000
	$51,568,000	$38,403,000
Less amounts due within one year	3,458,000	799,000
	$48,110,000	$37,604,000

At November 25, 1973 and November 26, 1972 the cost of properties pledged to secure indebtedness was $26,557,000 and $15,132,000, respectively.

The Company's principal note agreement, among other things, limits the declaration of dividends (other than stock dividends) and the redemption of its capital stock after April 1, 1971 to $12,000,000 plus consolidated net income, as defined, earned subsequent to November 29, 1970. Under this agreement, retained earnings at November 25, 1973, not so restricted, amount to $51,000,000. In addition, the Company may not declare dividends or redeem capital stock if its consolidated net current assets, as defined, would after such action be less than $45,000,000.

The Eurodollar note agreement, which was negotiated to satisfy requirements of the U.S. Office of Foreign Direct Investments, is a revolving credit agreement providing for borrowings of up to the equivalent of $10,000,000 in Eurodollars or foreign currencies at varying interest rates for renewable periods of from one to six months. Commencing on or about April 10, 1975, the maximum borrowings permitted under the agreement shall be reduced semi-annually in five equal installments. The Company had intended to continue borrowing the maximum allowable under the agreement until the mandatory repayment dates; therefore, the loan was classified in 1972 as long-term debt. However, changes in regulations governing foreign direct investments eliminated the need for continuous Eurodollar borrowings. In 1973 the amounts borrowed under this agreement were repaid.

The aggregate long-term debt maturities for the next five fiscal years are:

Fiscal Year	Principal Payments
1974	$3,458,000
1975	3,930,000
1976	2,810,000
1977	2,428,000
1978	3,363,000

SHORT-TERM DEBT AND LINES OF CREDIT

The Company and its subsidiaries had available at November 25, 1973 unused credit lines from domestic and foreign banks of $62,000,000 and $60,000,000, respectively. The domestic lines provide for borrowings on renewable short-term notes at bank prime interest rates. The foreign lines provide for borrowings on both renewable short-term notes and on an overdraft basis at varying rates which averaged 10% during the year. Combined domestic and foreign short-term borrowings for the year averaged $71,000,000 at an average interest rate of 9%.

Maximum short-term borrowings during the year were $91,000,000. At year end, short-term borrowings were all from foreign lenders at an average interest rate of 11.5%.

The terms of the credit lines with domestic banks generally include an informal arrangement as to average annual compensating balance requirements. Credit arrangements with foreign lenders normally have no such requirement.

INCOME TAXES

The provision for taxes on income consists of:

1973	Federal	State	Foreign	Total
Current	$14,305,000	$1,666,000	$4,753,000	$20,724,000
Deferred	(29,000)	—	1,257,000	1,228,000
	$14,276,000	$1,666,000	$6,010,000	$21,952,000
1972				
Current	$14,975,000	$1,505,000	$6,683,000	$23,163,000
Deferred	469,000	—	(586,000)	(117,000)
	$15,444,000	$1,505,000	$6,097,000	$23,046,000

The deferred income tax provision results from timing differences in the recognition of income and expense for tax purposes and for financial reporting purposes. The principal timing difference relates to the treatment of intercompany profits in inventories of foreign subsidiaries which are recognized currently for tax purposes and on the ultimate sale of the merchandise for financial reporting purposes. In 1972, intercompany profits for tax purposes exceeded those recognized for financial reporting purposes with a resulting credit to deferred taxes of $600,000. In 1973, the reverse occurred with a charge to deferred taxes of $1,070,000.

The provision for taxes on income for 1973 and 1972 as shown above differs from the amounts computed by applying the U.S. federal income tax rate (48%) to income before taxes. The principal reasons for this difference are:

	1973	1972
Tax computed at 48%	$16,228,000	$23,073,000
Increases (reductions) in taxes resulting from:		
Losses on foreign operations where the benefits cannot be currently utilized	7,103,000	91,000
State taxes, net of federal income tax benefit	862,000	771,000
Differences in income tax rates between the United States and foreign countries	(699,000)	(1,016,000)
Indefinite deferral of the tax liability on a portion of the DISC income	(566,000)	(136,000)
Flow through of investment tax credit	(420,000)	(520,000)
Other, net	(556,000)	783,000
Actual tax provision	$21,952,000	$23,046,000

At November 25, 1973, certain foreign subsidiaries had cumulative losses of approximately $14,500,000 which are available to reduce the future taxable income of those subsidiaries. Approximately $5,300,000 of these losses are available indefinitely and substantially all of the remainder are available for five years. These losses are subject to review and possible adjustment by the tax authorities of the countries involved.

STOCK OPTIONS

Under the Plan in effect, stock options to purchase up to 500,000 shares of the Company's common stock may be granted to eligible employees until November 1, 1980. Options may be granted at prices not less than 100% (qualified stock option) and 85% (non-qualified stock option) of the quoted market price on the date of the grant, subject to certain limitations. Such options become exercisable in cumulative installments of up to 20% during each of the second and third years and 30% during each of the fourth and fifth years after grant. Options expire five years after the date of the grant in the case of qualified options and ten years in the case of non-qualified options. The maximum number of shares for which options may be granted to any one employee under the Plan is 15,000 shares.

During 1972, options on 83,000 shares were granted at $56.38 per share but were cancelled in 1973 — none of the options had been exercised. During 1973 options were granted on 119,850 shares at $20.25 per share and on 2,500 shares at $23.00 per share. No options have been exercised.

At November 25, 1973 options were outstanding on 15,000 shares at $37.88, on 119,850 shares at $20.25, and on 2,500 shares at $23.00, leaving 362,650 shares available for subsequent grants under the Plan.

RETIREMENT PLANS

The Company and certain of its subsidiaries have non-contributory pension plans and contributory profit-sharing and factory savings plans which provide retirement benefits for their employees except those covered by union plans. At November 25, 1973, unfunded past service costs of the pension plans approximated $540,000 and the assets of those plans exceeded the actuarially computed value of vested benefits.

The Factory Savings Plan provides for voluntary contributions by participating employees up to 5% of earnings, not to exceed $600 for any employee annually. The Company is presently required to match employee contributions. Ultimately, the Company's contributions will range from one to three times the contributions by employees, depending upon the number of consecutive years each employee has been contributing.

Subject to certain limitations, the Profit-Sharing Plan requires minimum annual contributions of 1.95% of net income before provision for income taxes and contributions to the Plan. Participating employees may make voluntary contributions of up to 10% of their earnings, and may designate one-half or less of the contribution to be invested in Company stock. Such stock is purchased on the open market. Subject to the same limitations, the Company and the participating subsidiaries make an additional contribution of 50% of employee contributions designated for purchase of Company stock. The additional Company contribution is also invested in Company stock. In 1973, the Company did not attain a sufficient level of profit to permit any contribution to the Profit-Sharing Plan.

The aggregate cost of these plans for 1973 and 1972 totaled $1,368,000 and $2,355,000, respectively.

LEASES

The Company and its subsidiaries are obligated under long-term leases for real estate (office space, warehouses, plants and other facilities) and equipment, primarily truck fleet and computers. Rental expenses under such leases were $7,300,000 and $5,250,000 in 1973 and 1972, respectively.

At November 25, 1973, the minimum rental commitments under these leases were:

	Year Ending November					Five Years Ending November			Total
	1974	1975	1976	1977	1978	1983	1988	1993	
				(In Thousands of Dollars)					
Real estate	$7,190	$6,640	$5,910	$5,720	$5,430	$19,800	$15,320	$14,030	$80,040
Equipment	2,210	1,590	1,000	950	910	2,100	—	—	8,760
	$9,400	$8,230	$6,910	$6,670	$6,340	$21,900	$15,320	$14,030	$88,800

In general, the leases relating to real estate include renewal options for periods up to twenty years and escalation clauses relating to increases in the operating costs of the leased facilities.

The Company has no significant "financing leases" that have not been capitalized.

COMMITMENTS AND CONTINGENT LIABILITIES

The estimated cost to complete and equip construction and renovation projects in progress or committed for at November 25, 1973 is approximately $12,500,000.

In 1969, a defendant in a lawsuit instituted by the Company, filed a countersuit claiming unpaid royalties under a license agreement and damages for patent infringement and restraint of trade. This case has been heard but a decision has yet to be handed down. In management's opinion, based upon the advice of counsel, the decision will not have a material effect on the Company's financial position.

The Company has agreed with the trustees of the trusts established under the Profit-Sharing Plan and the Factory Savings Plan to purchase, if so requested, the Company's common stock held in trust while such stock is restricted against public sale. Such shares would be acquired at the market price. At November 25, 1973, 241,600 shares held in trust were so restricted.

REPORT OF INDEPENDENT PUBLIC ACCOUNTANTS
TO THE STOCKHOLDERS AND BOARD OF DIRECTORS OF LEVI STRAUSS & CO.:

We have examined the consolidated balance sheet of Levi Strauss & Co. (a Delaware corporation) and subsidiaries as of November 25, 1973, and the related consolidated statements of income, stockholders' equity, and changes in financial position for the year then ended. Our examination was made in accordance with generally accepted auditing standards, and accordingly included such tests of the accounting records and such other auditing procedures as we considered necessary in the circumstances. The consolidated balance sheet as of November 26, 1972, and the related consolidated statements of income, stockholders' equity and changes in financial position for the year then ended, which are presented for comparative purposes, were examined and reported on by other public accountants.

In our opinion, the accompanying financial statements present fairly the financial position of Levi Strauss & Co. and subsidiaries as of November 25, 1973, and the results of their operations and the changes in their financial position for the year then ended, in comformity with generally accepted accounting principles applied on a basis consistent with that of the prior period.

San Francisco, California.
January 15, 1974

Arthur Andersen & Co.

QUESTIONS

2-1. (a) What is meant by the term generally accepted accounting principles?
    (b) Are these principles ever subject to revision? Explain.

2-2. (a) What does a balance sheet prepared for a particular company present?
    (b) If XYZ Company prepares a balance sheet on December 31, 1974, for what period of time is the balance sheet a relevant representation of the financial position of XYZ Company? Explain.

2-3. Define the following:
    (a) Business entity assumption
    (b) Assets
    (c) Equities
    (d) Working capital
    (e) Current assets
    (f) Current liabilities

2-4. Of what value is the calculation of a current ratio for a particular company? Is the information gained by calculating the current ratio of value to anyone other than company management? In what way?

2-5. What is the order of presentation on the balance sheet of current assets and current liabilities?

2-6. Define a fixed asset and give examples of assets that would be classified as fixed.

2-7. What is the difference between liabilities and owners' equity?

2-8. Of what value is the calculation of the debt-equity ratio for a particular company? Is the information gained by calculating the debt-equity ratio of value to anyone other than company management? In what way?

2-9. (a) What does an income statement prepared for a particular company present?
    (b) Why is it advisable to present income statements for one or more previous years along with the current income statement?

2-10. Discuss the term *matching* as it applies to the preparation of the income statement.

2-11. Define the following:
    (a) Net sales
    (b) Cost of goods sold
    (c) Operating income
    (d) Net income

2-12. The income taxes paid by a particular business are a result of that business being in operation; however, income taxes are not classified as an operating expense on the income statement. What is the reason for this apparent discrepancy?

2-13. What is the purpose of calculating the rate of return on total assets?

2-14. What is the ratio for the rate of return on owners' equity? Of what value is the calculation?

2-15. What is the ratio for calculating earnings per share? Of what value is this calculation?

2-16. (a) Define retained earnings.

(b) What events cause the amount of retained earnings to change?

2-17. Why would a prospective investor want to compare the price-earnings ratio of various firms as opposed to the market price when deciding on a prospective investment?

PROBLEMS

Procedural Problems

2-1. The Sanderson Company had the following assets, liabilities, and capital accounts on December 31, 1974:

Cash	$ 17,500
Buildings	88,000
Accounts payable	16,800
Inventory	35,600
Land	82,000
Common stock	100,000
Marketable securities (Cost)	13,200
Accumulated depreciation—building	28,000
Patents	9,500
Notes payable due in 1977	24,600
Income tax payable	8,200
Retained earnings	33,200
Premium on common stock	30,000
Accounts receivable	27,000
Mortgage payable due in 1985	30,000
Allowance for uncollectible accounts	2,000

REQUIRED. Based upon the preceding accounts, prepare a balance sheet.

2-2. The Calumet Company has accumulated the following data pertaining to the company's operations for 1975:

Advertising expense	$ 8,000
Interest and dividend income	10,000
Salaries expense—salesmen	12,800
Rent expense	4,800
Insurance expense	3,000
Sales	203,500
Federal and state income taxes	20,000
Cost of goods sold	88,000
Sales returned by customers	3,500
Salaries expense—executives	30,000
Office supplies expense	2,000

REQUIRED
(a) Prepare an income statement for Calumet Company (omit earnings per share calculations).
(b) What is the average gross profit margin?

2-3. Using the data compiled in problem 2-1, compute the following:
(a) Working capital
(b) Current ratio
(c) Debt-equity ratio
Evaluate the financial condition of Sanderson Company based upon these ratios.

2-4. The Wendleford Company began operations on January 1, 1975. During the year the company accumulated the following information:

Selling expense	$ 6,000
Inventory	17,500
Building	35,000
Bonds payable	20,000
Cash	9,400
Notes payable — due 6/30/76	6,000
Cost of goods sold	56,000
Administrative expense	9,600
Sales	89,400
Copyrights	10,000
Accumulated depreciation — building	3,500
Accounts payable	8,300
Accounts receivable	12,500
Retained earnings	?
Common stock (10,000 shares)	50,000
Federal income tax expense	5,900
Land	20,000
General expense	2,500
Allowance for uncollectible accounts	1,300

REQUIRED. Using the information just given, prepare an income statement and balance sheet. The amount of retained earnings must be computed in order for the balance sheet to be prepared. Earnings per share should be computed.

2-5. Using the data compiled in Problem 2-4, compute the following (carry percentage answers to two decimal places):
(a) Working capital
(b) Current ratio
(c) Debt-equity ratio
(d) Rate of return on assets
(e) Rate of return on owners' equity
(f) Price-earnings ratio (assume market price of $15)
Based upon the preceding analysis, speculate as to whether or not you would invest in the stock of Wendleford Company.

2-6. The Stukel Corporation purchased a plot of land paying $15,000 on October 31, 1975. On November 15, 1975 the Corporation received an offer of

$18,000 for the land which they accepted. The firm wishing to purchase the land promised to pay the full amount on December 15, 1975. Title to the land was transferred on November 15.

REQUIRED. Explain the effect of the purchase and sale of the land on the balance sheet and income statement of Stukel Corporation assuming such statements were prepared in November 1975 and in December 1975.

Conceptual Problems

2-7. The Carter Company has current assets of $100,000 and current liabilities of $50,000. Thus, the working capital is $50,000 and the current ratio is 2:1.

REQUIRED. Consider each of the following events independently and explain how the working capital and current ratio of the Carter Company is affected by each. Use the terms increase, decrease, or no change. In answering the question, consider the effect on the individual assets and equities.

(a) Borrowed $2,000 from the bank to be repaid in six months.
(b) Purchased a machine which has an expected useful life of ten years, paying $8,000 cash.
(c) Sold $1,000 of common stock to a new investor for cash.
(d) Purchased land costing $10,000, giving a note payable due in ten years as payment.
(e) Collected a $600 account receivable.
(f) Purchased $5,000 of inventory paying $2,500 in cash and agreeing to pay the remainder in six months.
(g) Paid a supplier $1,000 on account.

2-8. The Spindleside Company has been in business for three years. During this period, the company has maintained an incomplete set of accounting records. The company recently applied for a bank loan, but the bank officers were unable to grant the loan due to the lack of formal financial records. The president of Spindleside Company asks you to prepare a balance sheet in an attempt to secure the loan. The following information is available:

Cash on hand	$ 75
Cash in bank	225
Cost of inventory on hand	4,700
Amount owed by customers (net of allowance of $500)	6,500
Building owned by Spindleside	37,000
Taxes owed to federal government	6,000
Amount owed to suppliers	10,000
Equipment owned by Spindleside	15,000
Mortgage owed on building — due in 10 years	25,000
Sales commissions owed to salesmen	7,000
Profit retained in business	500

The president informs you that all the stock was sold when the Company began operations. The par value is $10,000, but $15,000 was received from the sale.

2-9. Assume you have recently inherited $50,000 and are interested in investing these funds in the stock of some corporation. What facts and information about a company do you consider most important from a potential investor's point of view? Why?

2-10. In Chapter 1, the concepts of relevance, quantifiability, and verifiability were presented. With respect to the following accounts, indicate the types of information which are relevant and quantifiable in arriving at these particular account balances. Also, indicate how this information can be verified.

    (a) Cash   (e) Accumulated depreciation
    (b) Accounts receivable   (f) Accounts payable
    (c) Inventories   (g) Common stock
    (d) Fixed assets

2-11. Alice Howell recently decided to invest some funds in stock of a publicly traded corporation. She is aware that many investors use the earnings per share to analyze the company's potential. Ms. Howell proceeded to review a number of companies using EPS as her basic criterion. Of all the investment possibilities, Stable Corporation showed the largest increase in EPS over the past year. Because EPS rose from $2.00 per share to $4.60 per share, Ms. Howell invested all of her funds in Stable's common stock. What Ms. Howell failed to recognize is the fact that the large increase was due to a sale of long-term investments at a substantial profit. In subsequent years, EPS decreased to the $2.00 per share range.

Ms. Howell is now quite upset with the management of Stable Corporation. Even though the annual report she used in making her decision indicated the increase was due to the extraordinary sale, she contends that the company management committed fraud. Also, she believes the corporation is cheating her because the dividend they pay has consistently been more than a dollar per share less than the EPS.

REQUIRED. Comment on the accusations of Ms. Howell concerning the actions of the corporation. Do you believe she has grounds for legal action?

2-12. The president of the company for which you work as chief financial officer informs you that he is not totally satisfied with the balance sheet and income statement you have prepared for the year. His major complaint is that all the assets are shown at cost and, as such, do not project the true value of the company. He argues that investors would pay far more for the company's stock if they knew what the assets were really worth. He thus requests that you prepare a special financial report which will be sent to all shareholders using the true value of the items owned by the company.

REQUIRED. Draft a reply to the president's request indicating the problems you see in preparing the statement as requested.

3

THE MEASUREMENT
AND RECORDING OF
ACCOUNTING DATA

EDUCATIONAL OBJECTIVES

The material presented in this chapter is designed to achieve several educational objectives. These include:

1. An understanding of the phases in the accounting cycle with particular emphasis on the measurement and recording processes.
2. A knowledge of the fundamental accounting equation ($A = L + E$) and the effects of business transactions on the equation.
3. A working knowledge of some technical accounting terms such as debit, credit, accounts, journal, and ledger.
4. An appreciation of how the various steps in the accounting cycle achieve the desired end of published financial statements.

In order for accounting to serve as a means of communication for business it is important that certain fundamental conventions, or procedures, be established. Such conventions will be clearly man-made and therefore subject to change. Furthermore, they may be different in other countries but, within a given economy or society, the individuals involved in preparing accounting reports should understand and follow the generally accepted conventions. Otherwise, accounting would create confusion rather than being the source of information for financial decisions.

In this chapter and the succeeding one, those conventions which are a part of the American accounting framework will be presented in the context of the accounting cycle. The accounting cycle is the term used to indicate the processes involved in identifying, measuring, and communicating economic information. The basic phases of the cycle are summarized in the following outline:

1. Business transactions are analyzed.
2. The transactions are recorded in the journal.
3. Journal entries are posted to the ledger accounts.
4. A trial balance is prepared from the ledger accounts.
5. Accounts are reviewed and the necessary adjustments determined and recorded in the general journal.
6. Adjusting journal entries are posted to the ledger.
7. Financial statements are prepared from the adjusted ledger account balances.
8. Temporary accounts are closed.

The first four phases of the cycle are discussed in this chapter and the remaining phases discussed in Chapter 4.

THE ACCOUNTING EQUATION

A fundamental convention in accounting is that of double-entry bookkeeping. This system was first described by Luca Paciolo in 1494; however, he was not the inventor of double-entry. There is substantial evidence that such a system of bookkeeping was in use long before Paciolo described the system.

Double-entry bookkeeping can be looked upon as a scale which must have equal weight on each of the two sides in order for the scale to be balanced. Therefore, if the weight is increased or decreased on one side the same weight must be added to, or removed, from the other side. The more common explanation or illustration of double-entry bookkeeping is in the form of an equation. The fundamental accounting equation, as indicated in Chapter 2, is:

$$A = E$$

where

$$A = \text{Assets}$$
$$E = \text{Equities}$$

In its simplest form, the accounting equation at the inception of the Keeneland Company can be illustrated as:

$$\text{Assets} = \text{Equities}$$
$$\text{Cash, } \$10,000 = \text{Owners' equity, } \$10,000$$

This indicates that the company was started by the owners contributing $10,000.

Since financial statements typically draw attention to the difference between the claims by owners and the claims by creditors, the accounting equation can be expanded to take into consideration these differences and becomes:

$$A = L + C$$

where

$$L = \text{Creditor claims (liabilities)}$$
$$C = \text{Owners' claims (capital)}$$

If Keeneland acquired $1,000 of inventory to be paid for in 30 days, the equation would become:

$$\text{Assets} = \text{Liabilities} + \text{Capital}$$

$$\text{Cash } \$10,000 +$$
$$\text{Inventory } \$1,000 = \text{Accounts payable } \$1,000 +$$
$$\text{Owners' equity } \$10,000$$

The claims of the owners are often referred to as residual, since, in a liquidation, they do not have any claim on the assets until all claims of the creditors have been satisfied. Within that context, the accounting equation could be expressed as:

$$\text{Assets} - \text{Liabilities} = \text{Owners' equity (owners' claims)}$$

The residual nature of the owners' claims means that the amount of those claims may be increased or decreased depending upon the success or failure of the business. If the business is profitable, the owners' claims will be increased. On the other hand, if the business is not profitable, the claims of the owners will, necessarily, be decreased.

The success or failure of a business, which is the increase or decrease in the owners' claims, is reflected by the income statement. This statement provides a summary of the revenue and expense of the business for a period of time measured in accordance with generally accepted accounting principles.

Revenue

Revenue is the inflow of assets which results from the sale of goods or services in the normal course of business and is an increase in capital. The usual procedure is to record revenue when it is earned. In accounting, revenue is normally earned when a firm has completed its major obligation under a contract, i.e., when the goods are shipped or services performed. Consequently, revenue is often recognized before cash is collected. For example, when merchandise is sold to a customer on a charge account, revenue has been earned at the point of sale and should be recorded. On the other hand, if a fee for services is received prior to the services being rendered, revenue has not been earned and the business should record a liability in order to reflect its obligation to perform a service in the future.

In order to illustrate the effect of revenue on the accounting equation assume that merchandise costing $500 was sold, on account for $800. Continuing the Keeneland example:

		Assets	= Liabilities + Capital	
	Accounts		Accounts	Owners'
Cash	+ receivable	+ Inventory	= payable	+ equity
$10,000		+ $1,000	= $1,000	+ $10,000
	+ $800	− 500	=	+ 300
$10,000	+ $800	+ $ 500	= $1,000	+ $10,300

Note that the accounts receivable increased by $800 and the inventory was reduced by $500. The net effect of this transaction was an increase of $300 in the assets which increased the owners' equity.

Expenses

Expenses are costs incurred in obtaining revenues which result in a decrease in assets or an increase in liabilities. Some common expenses are:

the cost of merchandise sold, employee salaries, building rent, usage of fixed assets, insurance, utilities, and taxes. The effect of expenses on the accounting equation can be illustrated by assuming that (a) salaries of $100 and (b) rent of $75 were paid:

	Cash	+	Accounts receivable	+	Inventory	=	Accounts payable	+	Owners' equity
	$10,000	+	$800	+	$500	=	$1,000	+	$10,300
(a)	−100					=		−	100
(b)	− 75					=		−	75
	$ 9,825	+	$800	+	$500	=	$1,000	+	$10,125

In order to have a proper matching of revenues and expenses, it is especially important to determine which costs relate to the revenues which have been recorded. If there is a postponement in the recording of revenue, then all costs incurred in connection with the revenue should be postponed and recorded as an asset rather than as an expense. Subsequently, when the revenue is recorded, the costs should be recorded as an expense and the asset reduced.

Effect of Revenue and Expense on the Accounting Equation

In the context of the equation Assets = Liabilities + Capital, revenue is an increase in capital and expense a decrease in capital. When revenue is earned, the claims of the owners are increased. When expenses are incurred, the claims of owners are reduced, and the amount of capital is decreased.

To increase or decrease owners' equity with each revenue and expense item could produce confusion relative to the total owners' equity since the total would be constantly changing. In addition, much useful information about the source of the increases and decreases would be lost by summarizing revenues and expenses into capital. Therefore, on a temporary basis, revenues and expenses are recorded in separate categories from the owners' equity, or capital. This convention gives rise to another expansion of the basic accounting equation which is expressed as:

$$A = L + [C + (R - E)]$$

At the end of each accounting period, the revenues (R) and expenses (E) are transferred to capital as a net amount. The excess of the revenues over the expenses is called *net income*. An excess of expenses over revenues is a *net loss*. As a result of classifying the increases in capital as revenues and the decreases as expenses, there is a detailed record of the sources of the change in capital, resulting from operations. Figure 3-1 illustrates the relationship between revenues, expenses, owners' withdrawal, and capital.

ACCOUNTS

In the preceding discussion, only the major components of the accounting equation were considered. In practice, however, there are several classifications of items within each of the major categories. The record of these individual classifications is called an *account*. An account is simply a place where similar transactions and events which occur throughout the period are summarized and accumulated. Transactions refer to business events which can be expressed in money terms and, therefore, must be recorded in the accounting records. In order to generate the maximum amount of information, the account must be designed to reflect the increases and decreases in a particular classification. In accounting, the increases and decreases are referred to as debits and credits.

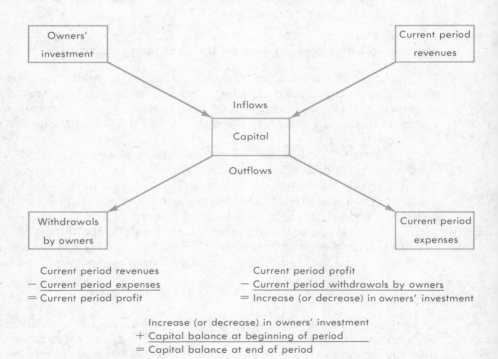

Current period revenues
− Current period expenses
= Current period profit

Current period profit
− Current period withdrawals by owners
= Increase (or decrease) in owners' investment

Increase (or decrease) in owners' investment
+ Capital balance at beginning of period
= Capital balance at end of period

FIGURE 3-1

The terms, *debit* and *credit,* refer to the accounting explanation of the mathematical changes in the individual accounts. In order for the accounting equation to remain in balance, there must be equal increases and decreases. Therefore, when there are debits there must be credits in an amount equal to the debits.

This accounting convention of a double-entry system simply indicates that the debits must be equal to the credits. The terms debit and credit do not mean good or bad and there is no special reason why the words debit and credit are used except that these terms are a part of accounting convention established over time.

Account Design

While the physical design of an account may vary from one accounting system to another, it is possible and convenient for illustrative purposes to use an account in the form of a T as shown here:

TITLE OF ACCOUNT

DEBIT	CREDIT

Many accountants, in analyzing transactions, make use of T accounts for computational purposes. Thus, T accounts are not formal accounting records or procedures, but they enable the accountant to analyze the effect of transactions on the accounts without using formal records. Each account, and consequently the T, has debits on the left, credits on the right, and the account title.

Accounts may be either manual or automated. The use of punched cards or magnetic tapes in the accounting system does not change the basic purpose of accounts. In a punched card system, the usual practice is to keypunch a card for each transaction. The columns of the card are designated for certain data such as date, account affected (expressed as a number), reference or source of transaction, and amount. (There may be one location for debits and another for credits or simply a special punch which signifies credit.)

A manual account may be designed either in a two-column account form or a balance-column account form. The two-column account form, which is illustrated and cross-referenced in Figure 3-2, usually contains the following information:

1. Name of account. That is, cash, accounts receivable, building, etc.

TWO-COLUMN ACCOUNT FORM

ACCOUNT: Cash[1]   [2]ACCOUNT NO. 101

DATE[3]		EXPLANATION[4]	POST[5] REF.	DEBIT[6]	DATE[3]		EXPLANATION[4]	POST[5] REF.	CREDIT[7]
1974 Oct.	1	Balance		5000	Oct.	20			10000
	15			12000					

BALANCE-COLUMN ACCOUNT FORM

ACCOUNT Cash   ACCOUNT NO. 101

	DATE 1974		EXPLANATION	POST. REF.	DEBIT	CREDIT	BALANCE DEBIT	BALANCE CREDIT	
1	Oct.	1	Balance				5000		1
2		15			12000		17000		2
3		20				10000	7000		3
4									4

FIGURE 3-2

2. Account number. A numerical designation for the account name. This is discussed in a later section of this chapter.

3. Date of recording the information. That is, October 15, 1974. A common practice is to indicate the year at the top of the column.

4. Explanation. This column is available for any notation considered relevant for informational purposes.

5. Posting reference. This column serves as a control device and will be discussed later in the chapter.

6. Debit. Any debits to the account are recorded in this column.

7. Columns for recording credits to the accounts are provided so that the same information provided for the debits is available for any credits.

In the balance-column account form similar information is available. The major difference is that a balance column is provided and the balance in the account is available at all times. In addition, there is only one set of columns for date, explanation, and posting reference.

Normal Balance. As a part of the double-entry convention, an increase in an asset is a debit. Therefore, assets usually have a debit balance. Liabilities and owners' equity, on the other hand, normally have credit balances because, according to the double-entry convention, they are increased by credits. Thus, all items found on the left side of the equation, $A = L + C$, normally have debit balances, and all items on the right have credit balances.

When the accounting equation was expanded to include revenue and expense it was necessary to have a negative E for the expenses. In order to eliminate a negative sign in the equation, the negative E is algebraically transferred to the left side of the equation and becomes positive. Consequently, within the framework of debits and credits, expenses have debit balances and would be on the left of the equation. The equation can now be stated:

$$A + E = L + C + R$$

An understanding of the normal balance of each item is important in order to know when to debit or credit the item. In order to increase assets or expenses, it is necessary to debit the particular asset or expense since a debit balance plus a debit will result in a larger debit balance. A reduction in assets or expenses requires a credit because a debit balance minus a credit results in a smaller debit balance. Conversely, liabilities, capital, and revenue are increased by credits since a credit balance plus a credit results in a larger credit balance. These items are decreased by a debit because a credit balance minus a debit results in a smaller credit balance.

The debit and credit convention of double-entry bookkeeping may be summarized as follows:

ASSETS—Debits increase; credits decrease; normally debit balance.

EXPENSES—Debits increase; credits decrease; normally debit balance.

LIABILITIES—Credits increase; debits decrease; normally credit balance.

CAPITAL—Credits increase; debits decrease; normally credit balance.

REVENUES—Credits increase; debits decrease; normally credit balance.

The cash account in the form of a T account will be used to illustrate debit balance accounts. Since cash is an asset account it normally has a debit balance. The receipt of cash represents an increase, or debit, of the cash account and cash disbursements are a decrease, or credit. In the following example, assume that the cash account had an opening balance of $5,000, and that during the period cash receipts were $12,000 and cash disbursements $10,000. The new balance is determined by adding the receipts, or debits, to the beginning balance and subtracting the disbursements, or credits.

CASH

COMPUTATIONS		DEBIT		CREDIT
Opening balance	$ 5,000	Opening balance	5,000	10,000 Disbursements
+ Receipts	12,000	Receipts	12,000	
	$17,000			
− Disbursements	10,000			
Ending balance	$ 7,000	Ending balance	7,000	

The use of T accounts for analyzing accounts with credit balances, i.e., liabilities, owners' equity, and revenues, will be illustrated by a T account for the accounts payable account. Assume that accounts payable had an opening balance of $7,000, and that the payments on account during the period were $25,000 and the merchandise purchased on account was $20,000. The new balance is determined by adding the purchases, or credits, to the opening balance and subtracting the payments, or debits.

ACCOUNTS PAYABLE

COMPUTATIONS		DEBIT	CREDIT
Opening balance	$ 7,000	Payments 25,000	7,000 Opening balance
+ Purchases	20,000		20,000 Purchases
	$27,000		
− Payments	25,000		
Ending balance	$ 2,000		2,000 Ending balance

Account Titles

The number of accounts and their titles are determined by individual firms; however, in most cases there will be accounts within each category of the expanded equation $A = L + [C + (R − E)]$. Accordingly, there will be one or more accounts within each of the following categories: Assets, Liabilities, Owners' Equity, Revenues, and Expenses. The greater the number of accounts, the more information provided to management relative to the operation of the business; however, the greater the number of accounts, the more complex the accounting process. For example, it may be relatively easy to determine that an expense has been incurred; however, if the company has 25 expense accounts it may be more difficult to determine the particular account to be debited for the expense.

Managerial Use of the System of Accounts

The design of the accounting system should take into consideration both the complexity of the company and the managerial need for relevant information. The cost of collecting information should never exceed its usefulness to management or other users. A large firm with diversified products and sales may need considerably more information than a small firm in order to evaluate the results of operation. A major point which should be remembered is that even though the accounting system has been expanded, the accounting equation must remain in balance.

CHART OF ACCOUNTS

In addition to the account titles, it is fairly common, especially in larger accounting systems, to give each account a number. Within a given business, the authorized accounts are referred to as the *chart of accounts*. Typically, the accounts are arranged, and have numbers assigned to them, in the order in which they appear in the balance sheet and income statement. For example, since assets appear first in the balance sheet, they would be assigned the lower numbers. In addition, assets are numbered based on their liquidity, i.e., nearness to cash. As an illustration of a chart of accounts, a three-digit numbering system will be used. The account numbering in the illustration is as follows:

100–199	Asset accounts	400–499	Revenue accounts
200–299	Liability accounts	500–599	Expense accounts
300–399	Capital accounts		

BALANCE SHEET ACCOUNTS

100   Assets
    101   Cash
    110   Accounts receivable
    111   Allowance for uncollectible accounts
    121   Supplies
    131   Merchandise inventory
    141   Prepaid insurance
    161   Buildings
    171   Accumulated depreciation — buildings
200   Liabilities
    201   Accounts payable
    211   Salaries payable
    221   Taxes payable
    251   Bonds payable
300   Capital
    301   Common stock
    351   Retained earnings

INCOME STATEMENT ACCOUNTS

400   Revenue
    401   Sales
    411   Service revenue
    451   Sales returns and allowances

500   Expenses
    501   Cost of goods sold
    531   Salary expense
    541   Supplies expense
    551   Depreciation expense
    561   Tax expense
    571   Insurance expense
    581   Bad Debt expense
    591   Miscellaneous expense

The chart of accounts illustrated is numbered in such a way that it can be easily expanded. For example, if management wants salary expense to be shown separately for Executive Salary, Office Salary, and Sales Salary, it is a simple matter to assign the numbers 531, 532, and 533, respectively, to those accounts.

GENERAL LEDGER

The individual accounts are normally maintained in something referred to as the *general ledger*. The general ledger may consist of a looseleaf binder in which individual account pages are inserted, a group of account cards placed in a file cabinet, a set of punched cards, or the memory unit of a computer. Thus, there is no special physical characteristic of a general ledger. In accounting, frequent reference is made to the general ledger, not because of its physical appearance but rather because the general ledger contains all the accounts and, consequently, provides a permanent record of the financial transactions of the firm.

Trial Balance

In order to insure that the general ledger is in balance, i.e., total debits equal total credits, a trial balance is prepared prior to the preparation of financial statements. A trial balance is a listing of all account balances from the general ledger. A formal trial balance may include the title of each account as well as the debit or credit balance. The trial balance of the XYZ Company is presented for illustrative purposes.

XYZ Company
Trial Balance
October 31, 1974

	Debit	Credit
Cash	$ 15,900	
Accounts receivable	30,230	
Supplies	1,800	
Merchandise inventory	5,000	
Prepaid insurance	700	
Buildings	82,000	
Accumulated depreciation— buildings		$ 12,000
Accounts payable		8,600
Salaries payable		1,500
Taxes payable		500
Bonds payable		17,000
Common stock		76,000
Retained earnings		12,250
Sales		98,100
Service revenue		5,060
Sales returns and allowances	2,405	
Cost of goods sold	62,775	
Salary expense	21,000	
Supplies expense	1,700	
Depreciation expense	3,000	
Tax expense	800	
Insurance expense	500	
Miscellaneous expense	3,200	
	$231,010	$231,010

A primary purpose of a trial balance is to determine if the total debits and credits from the ledger are in balance, and this may be accomplished by simply preparing an adding machine tape of all debit balances and another tape of all credit balances. A trial balance does not prove that all transactions were recorded in the proper accounts. For example, a supplies expense could have been recorded in the salary expense account. Thus, in the trial balance, debits would equal credits but the accounts would have incorrect balances.

GENERAL LEDGER

JOURNALS

As transactions occur it is possible to record them directly in the proper accounts as in the previous illustrations; however, such a procedure would become cumbersome and confusing if there were many transactions. Furthermore, after several recordings it would become difficult to locate a particular transaction. In order to facilitate the availability of information and to provide a chronological record of transactions, most accounting systems make use of one or more journals.

A *journal* is known as the book of original entry. This means that as transactions occur they are recorded first in the journal. The amounts shown in the journal are then recorded in the individual accounts. Since the transactions are journalized in chronological order, the task of later tracing some particular transaction is simplified.

The form of the journal may be similar to the one illustrated which has a column for the date, description or account title, posting reference, debit, and credit.

A typical general journal contains the following information (see Figure 3-3):

1. The title *General Journal* is used to distinguish this journal from special journals which will be discussed in a later section. As the title indicates, the journal is used for all transactions which are not recorded in special journals. In addition, in this text all journal entries will be recorded in general journal form unless otherwise indicated.

2. Page number. This is the page in the journal, and it serves as posting reference in the ledger account, which was previously discussed.

	DATE 1974[3]		DESCRIPTION[4]	POST. REF.[5]	DEBIT[6]	CREDIT[7]	
1	Jan	12	Cash	101	2000		1
2			Sales	401		2000	2
3							3
4		14	Salary Expense	531	500		4
5			Cash	101		500	5
6							6

GENERAL JOURNAL[1]   PAGE 1[2]

FIGURE 3-3

3. Date. As transactions occur they are recorded in chronological order.

4. Description. The account titles which are involved in the transaction are shown in this column. The account(s) to be debited is shown first and placed next to the left margin. The account(s) to be credited is indented, as shown in the illustration. If the entry does not clearly identify the nature of the transaction, or if additional information concerning the transaction is desired, an explanation is written below the account titles. For example, the explanation of the entry of January 14, 1974 would be written as "salaries for the period ending 1/12/74."

5. Posting reference. Transactions are recorded in the journal as they occur. Then periodically, which may be daily, weekly, or monthly, the amounts are posted to the appropriate accounts. As each amount is posted, it is common practice to indicate in the ledger account the page number of the journal from which the entry originated. On the journal page, a check mark or the account number is placed in the posting reference column to indicate that the item has been posted to the ledger. Thus, posting means that the information in the journal has been transcribed into the appropriate ledger account.

6. Debit. A column for the debit amount is provided. There may be one or more debit accounts, and, therefore, debit amounts.

7. Credit. A column for the credit amount is provided. There may be one or more credit accounts and, therefore credit amounts.

The relationship between the general journal and the ledger account is illustrated in the posting of the journal entries to the appropriate accounts. In Figure 3-4, the balance-column account form is used.

SPECIAL JOURNALS

When a firm has many similar transactions, a common practice is to make use of special journals, in addition to the general journal. The more common special journals are sales journal, cash receipts journal, cash disbursements journal, and accounts payable or purchases journal. These journals serve exactly the same purpose as the general journal, i.e., a recording of the debits and credits of a transaction.

The use of special journals minimizes the time and effort required to record and post repetitive transactions because they are designed with special columns to summarize the data. The total of each column, which would probably reflect one month's activity, is often all that must be posted to the ledger account.

ACCOUNT Cash   **ACCOUNT NO.** 101

	DATE	EXPLANATION	POST. REF.	DEBIT	CREDIT	BALANCE DEBIT	BALANCE CREDIT	
1	Jan. 12		GJ1	2000		2000		1
2	14		GJ1		500	1500		2
3								3

ACCOUNT Sales   **ACCOUNT NO.** 401

	DATE	EXPLANATION	POST. REF.	DEBIT	CREDIT	BALANCE DEBIT	BALANCE CREDIT	
1	Jan. 12		GJ1		2000		2000	1
2								2

ACCOUNT Salary Expense   **ACCOUNT NO.** 531

	DATE	EXPLANATION	POST. REF.	DEBIT	CREDIT	BALANCE DEBIT	BALANCE CREDIT	
1	Jan. 14		GJ1	500		500		1
2								2

FIGURE 3-4

In addition to saving time and effort, special journals reduce the number of errors which can result from recording and posting numerous repetitive transactions. Similar transactions are located in a certain journal; therefore, if the nature of the transaction is known, its proper location can be determined without looking through all transactions. For example, the recording of a cash payment from a customer will be recorded in the cash receipts journal. If a customer inquires about his account, verification of cash received can be determined by reviewing the cash receipts journal rather than all transactions in the general journal.

Sales Journal

A sales journal serves to collect, in one central location, information relative to all credit sales. Since credit sales are common in most firms, the recording of individual sales in the general journal would be time-consuming both in terms of recording and posting to the ledger accounts.

Figure 3-5 illustrates the common format of a sales journal. When sales are made on credit there is an increase in the asset, Accounts Receivable,

SALES JOURNAL

DATE	INVOICE NUMBER		POST. REF.	ACCTS. REC. DR. SALES CR.
Mar. 1	201	Kern, Inc.	✓	562.40
5	202	Miles Co.	✓	1,715.05
15	203	Perry Distributors	✓	872.00
21	204	Grayson Co.	✓	2,370.10
28	205	Boone, Inc.	✓	722.50
31				6,242.05
				(110) (401)

GENERAL LEDGER ACCOUNTS

Accounts Receivable   No. 110

DATE	POST. REF.	DEBIT	CREDIT	BALANCE	
				DEBIT	CREDIT
Mar. 31	SJ6	6,242.05		6,242.05	

Sales   No. 401

DATE	POST. REF.	DEBIT	CREDIT	BALANCE	
				DEBIT	CREDIT
Mar. 31	SJ6		6,242.05		6,242.05

FIGURE 3-5

and a corresponding increase in the revenue, Sales. Therefore, since the recording is always the same, it is possible to have one column headed Accounts Receivable and Sales. The total of the column, which generally represents the sales for one month, is posted to both accounts.

The customer name is included in the sales journal because the firm must have a record of what each customer owes. These individual records are maintained in what is known as a *subsidiary ledger*. The format of subsidiary ledger accounts is the same as general ledger accounts except that customer names are used as the account titles. The numbers in parentheses below the monthly column total indicate that the total amount has been posted to those accounts. In this example, the total was

SPECIAL JOURNALS

posted to the Accounts Receivable (110) and Sales (401) accounts. A check mark in the posting reference column indicates that the transaction amount has been posted to the customer's account.

Subsidiary records are not the results of using special journals. These records are necessary even if the transactions are recorded in the general journal. An important point, relative to subsidiary ledgers, is that the total of the individual accounts must always be equal to the balance in the general ledger account. The general ledger account is called the *controlling account* and is the summary of the subsidiary ledger accounts.

Cash Receipts Journal

The journal illustrated in Figure 3-6 is referred to as a cash receipts journal because every transaction recorded in it increases the cash account. Consequently, there is a column headed Cash Debit and the total of that column indicates the amount of cash received by the firm during the period from all sources.

The most common sources of cash receipts are cash sales and collection of accounts receivable. The journal illustrated includes columns for Sales Credit and Accounts Receivable Credit, indicating the frequency of credits to these accounts. If cash is received from any other source the amount is entered in the Miscellaneous Credit column. The explanation column reflects the account to be credited for the miscellaneous receipts.

The posting of the cash receipts journal is similar to the sales journal. The column totals may be posted for Sales, Accounts Receivable, Sales

CASH RECEIPTS JOURNAL						PAGE 7
DATE	ACCTS. CREDITED EXPLANATION	POST. REF.	MISC. CR.	SALES CR.	ACCTS. REC. CR.	CASH DR.
Mar. 1	Cash sales	✓		267.50		267.50
8	Kern, Inc.	✓			562.40	562.40
17	Miles Co.	✓			1,715.05	1,715.05
22	Perry Distributors	✓			872.00	872.00
30	Notes receivable	115	500.00			525.00
	Interest income	460	25.00			
31			525.00	267.50	3,149.45	3,941.95
			✓	(401)	(110)	(101)

FIGURE 3-6

ACCOUNT Cash   ACCOUNT NO. 101

	DATE	EXPLANATION	POST. REF.	DEBIT	CREDIT	BALANCE DEBIT	BALANCE CREDIT	
1	Mar. 1		Bal			5000		1
2	31		CR7	394195		894195		2
3								3

ACCOUNT Accounts Receivable   ACCOUNT NO. 110

	DATE	EXPLANATION	POST. REF.	DEBIT	CREDIT	BALANCE DEBIT	BALANCE CREDIT	
1	Mar. 31		SJ6	624205		624205		1
2	31		CR7		314945	309260		2
3								3

ACCOUNT Notes Receivable   ACCOUNT NO. 115

	DATE	EXPLANATION	POST. REF.	DEBIT	CREDIT	BALANCE DEBIT	BALANCE CREDIT	
1	Mar. 1		Bal			500		1
2	30		CR7		500	0		2
3								3

ACCOUNT Sales   ACCOUNT NO. 401

	DATE	EXPLANATION	POST. REF.	DEBIT	CREDIT	BALANCE DEBIT	BALANCE CREDIT	
1	Mar. 31		SJ6		624205		624205	1
2	31		CR7		26750		650955	2
3								3

ACCOUNT Interest Income   ACCOUNT NO. 460

	DATE	EXPLANATION	POST. REF.	DEBIT	CREDIT	BALANCE DEBIT	BALANCE CREDIT	
1	Mar. 30		CR7		2500		2500	1
2								2

FIGURE 3-6 Continued

Discounts, and Cash. Each item in the miscellaneous column must be posted to the appropriate account. For the miscellaneous items, the account number is used as the posting reference. In addition, each collection of an account receivable must be posted to the individual customer's account in the subsidiary ledger.

Purchases Journal

The purchases journal, which may be referred to as the *accounts payable journal*, contains information concerning the acquisitions made on credit. As may be seen in Figure 3-7, everything recorded in this journal increases Accounts Payable. The most common purchases are merchandise and supplies; therefore, columns are normally provided for these two types of purchase. If there are other repetitive purchases another column may be added. The miscellaneous debits section of the journal is for acquisitions other than merchandise and supplies.

In the case of purchases on credit, a record of what the firm owes to individual creditors must be maintained as well as the total amount in Accounts Payable. These records are the subsidiary ledger accounts for accounts payable. The check mark in the posting reference column indicates that the amount has been posted to the individual account.

Cash Payments Journal

The cash payments journal, shown in Figure 3-8, is the record of all cash disbursements. The column headed Cash Credit reflects all payments made during the period. The columns for Accounts Payable and Merchandise Inventory Debit indicate that these accounts are generally involved in cash payments. As in other special journals, additional columns may be added if there are other common cash payments. The posting of column totals is the same as with the other journals, and the transactions, shown in the miscellaneous column, must be individually posted to the proper account. All debits to Accounts Payable must also be posted to the individual creditor's account.

A NOTE ON TECHNICAL PROCEDURES

Many mechanical and technical procedures have been presented in this chapter. There are several reasons for this:

1. The procedures are an essential component of the accounting processes.
2. The accuracy of the mechanical procedures has a significant impact on the validity of the accounting information.

			PURCHASES JOURNAL			PAGE 12		
DATE	ACCOUNT CREDITED	POST. REF.	ACCTS. PAY. CR.	MERCHANDISE INVENTORY DR.	SUPPLIES DR.	MISC. DEBITS		
						ACCT.	POST. REF.	AMOUNT
Mar. 1	Williams Bros.	✓	928.50	928.50				
7	Miller Co.	✓	1,570.00	1,570.00				
16	Young Distributors	✓	1,245.00	1,245.00				
20	Edwards Inc.	✓	530.50			Office equip.	181	530.50
25	Williams Bros.	✓	1,305.00	1,305.00				
30	City Supply Co.	✓	215.00		215.00			
31			5,794.00	5,048.50	215.00			530.50
			(201)	(131)	(121)			✓

ACCOUNT Supplies   ACCOUNT NO. 121

	DATE	EXPLANATION	POST. REF.	DEBIT	CREDIT	BALANCE		
						DEBIT	CREDIT	
1	Mar. 31		PJ12	215 00		215 00		1
2								2

ACCOUNT Merchandise Inventory   ACCOUNT NO. 131

	DATE	EXPLANATION	POST. REF.	DEBIT	CREDIT	BALANCE		
						DEBIT	CREDIT	
1	Mar. 31		PJ12	5048 50		5048 50		1
2								2

ACCOUNT Office Equipment   ACCOUNT NO. 181

	DATE	EXPLANATION	POST. REF.	DEBIT	CREDIT	BALANCE		
						DEBIT	CREDIT	
1	Mar. 31		PJ12	530 50		530 50		1
2								2

ACCOUNT Accounts Payable   ACCOUNT NO. 201

	DATE	EXPLANATION	POST. REF.	DEBIT	CREDIT	BALANCE		
						DEBIT	CREDIT	
1	Mar. 31		PJ12		5794 00		5794 00	1
2								2

FIGURE 3-7

CASH PAYMENTS JOURNAL

DATE	CK. NO.	ACCOUNT DEBITED	POST. REF.	MISC. DR.	ACCTS. PAY. DR.	MERCHANDISE INVENTORY DR.	CASH CR.
Mar. 2	229	Cash purchase	✔			113.00	113.00
9	230	Williams Bros.	✔		928.50		928.50
15	231	Salary expense	531	2,100.00			2,100.00
16	232	Miller Co.	✔		1,570.00		1,570.00
24	233	Prepaid insurance	141	315.00			315.00
31	234	Salary expense	531	2,250.00			2,250.00
31				4,665.00	2,498.50	113.00	7,276.50
				✔	(201)	(131)	(101)

ACCOUNT Cash   ACCOUNT NO. 101

	DATE	EXPLANATION	POST. REF.	DEBIT	CREDIT	BALANCE DEBIT	BALANCE CREDIT	
1	Mar. 1		Bal.			500000		1
2			CR7	394195		894195		2
3			CP11		727650	166545		3
4								4

ACCOUNT Merchandise Inventory   ACCOUNT NO. 131

	DATE	EXPLANATION	POST. REF.	DEBIT	CREDIT	BALANCE DEBIT	BALANCE CREDIT	
1	Mar. 31		PJ12	504850		504850		1
2	31		CP11	11300		516150		2
3								3

ACCOUNT Prepaid Insurance   ACCOUNT NO. 141

	DATE	EXPLANATION	POST. REF.	DEBIT	CREDIT	BALANCE DEBIT	BALANCE CREDIT	
1	Mar. 24		CP11	31500		31500		1
2								2

FIGURE 3-8

ACCOUNT Accounts Payable   ACCOUNT NO. 201

| | DATE | | EXPLANATION | POST. REF. | DEBIT | CREDIT | BALANCE | | |
							DEBIT	CREDIT	
1	Mar.	31		PJ12		579400		579400	1
2		31		CP11	249850			329550	2
3									3

ACCOUNT Salary Expense   ACCOUNT NO. 531

| | DATE | | EXPLANATION | POST. REF. | DEBIT | CREDIT | BALANCE | | |
							DEBIT	CREDIT	
1	Mar.	15		CP11	210000		210000		1
2		31		CP11	225000		435000		2
3									3

FIGURE 3-8 Continued

3. An understanding of the procedures enhances one's ability to understand and use accounting information.

4. These procedures will be used and alluded to throughout the remainder of the text.

Thus the procedures are necessary but should not be mistaken for the substance of accounting. Remember that the purpose of accounting is to identify, measure, and communicate useful economic information. In order to accomplish these goals, certain mechanical procedures are necessary.

SUMMARY

The first steps in the accounting cycle have been presented in this chapter. Those transactions identified as accounting information were recorded chronologically in the journal reflecting the debit and credit aspects of the transaction. Therefore, the journal provides a permanent record of all financial transactions of the firm.

In order to effectively communicate financial information to users, the numerous transactions must be summarized. Unless the information is summarized the user would have to review the entire journal to make decisions relative to the firm. This would be time-consuming and would not serve one of the purposes of accounting which is communicating useful economic information.

The summarization of information is by major classifications of assets, liabilities, and owners' equity. These items are further classified according to the chart of accounts. The design of the classification system is a particularly important feature relative to the communication of useful information.

In this chapter, several conventions which affect the accounting cycle have been presented. Among the more significant conventions are:

1. Double-entry bookkeeping is fundamental.
2. The basic accounting equation, Assets = Equities, is an expression of double-entry bookkeeping.
3. The basic equation may be expanded to: Assets = Liabilities + [Capital + (Revenue − Expense)].
4. Asset and Expense accounts are increased by debits and normally have a debit balance.
5. Liability, Capital, and Revenue accounts are increased by credits and normally have a credit balance.

In addition to presenting certain accounting conventions, this chapter has illustrated the first four phases of the accounting cycle:

1. Analyzing business transactions.
2. Recording transactions in a journal.
3. Posting journals to the ledger accounts.
4. Preparing a trial balance.

The remaining phases of the cycle will be discussed in the following chapter.

QUESTIONS

3-1. What are the several basic phases of the accounting cycle?

3-2. (a) What is the definition of the term assets as used in accounting?
    (b) List several examples of assets.

3-3. (a) What are equities?
    (b) List several examples of equities.
    (c) List the two divisions of equities and define each.

3-4. (a) State the fundamental accounting equation.
    (b) Expand the equation in (a) to indicate the difference between creditors' equity and owners' equity.
    (c) Rearrange the equation in (b) to show owners' equity to be a residual claim.

3-5. What is the purpose of the income statement?

3-6. (a) What is revenue and when is it normally recorded?
    (b) What are expenses and when should they be recorded?
    (c) Why are revenues and expenses temporarily recorded in separate categories from capital?

3-7. Using the accounting convention of debit and credit, describe how an account is increased and decreased, and state its normal balance by completing the following chart:

Type of account	Increase	Decrease	Normal balance
(a) Assets	_____	_____	_____
(b) Liabilities	_____	_____	_____
(c) Capital	_____	_____	_____
(d) Revenues	_____	_____	_____
(e) Expenses	_____	_____	_____

3-8. (a) What is an account?

    (b) What is a chart of accounts? How is it arranged?

3-9. Classify each of the following accounts as (a) Assets, (b) Liabilities, (c) Capital, (d) Revenues, or (e) Expenses:

    (1) Cash

    (2) Merchandise inventory

    (3) Accounts payable

    (4) Common stock

    (5) Sales

    (6) Salaries payable

    (7) Salaries expense

    (8) Accounts receivable

    (9) Retained earnings

    (10) Buildings

3-10. (a) "The general ledger is the same thing as the chart of accounts." Is this statement true or false? Why?

    (b) What is the device used by accountants to insure that the general ledger is in balance? Describe.

3-11. (a) What is a journal and why is it known as the book of original entry?

    (b) What is meant by the term posting as used in accounting? How often is the posting process performed?

    (c) Name and explain the technique used to insure that all journal entries are posted to the general ledger.

3-12. (a) Give three reasons why special journals should be used.

    (b) What types of transactions should be recorded in each of the following journals?

    (1) Sales journal

    (2) Cash receipts journal

    (3) Purchases journal

    (4) Cash payments journal

    (5) General journal

PROBLEMS

Procedural Problems

3-1. On December 31, 1974, the end of its fiscal year, the Hager Company had the following accounts in its general ledger—all with normal balances:

Account	Balance
Accounts payable	$ 16,900
Accounts receivable	34,000
Buildings	76,000
Cash	25,000
Common stock	75,000
Insurance expense	3,000
Merchandise inventory	45,000
Miscellaneous expense	1,400
Prepaid insurance	8,000
Cost of goods sold	80,200
Retained earnings	25,000
Salaries payable	6,000
Salary expense	60,000
Sales	220,000
Supplies	5,000
Supplies expense	5,300
Tax expense	17,000
Taxes payable	17,000

REQUIRED. Prove that Hager Company's general ledger is in balance by preparing a trial balance arranging the accounts in the order they should appear on financial statements.

3-2. As of December 31, 1974, the general ledger of the Kinnison Kwik Service Company had the following balances:

Cash	$20,750
Accounts receivable	10,650
Merchandise inventory	12,000
Prepaid insurance	600
Office equipment	8,000
Accounts payable	9,300
Salaries payable	200
Taxes payable	150
Common stock	25,000
Retained earnings	17,350

Transactions completed during January, 1975 included:

Jan. 2 Paid salaries owed at December 31, 1974.
       4 Purchased merchandise on account, $2,600.
       5 Cash sales for the week, $4,000.
       8 Borrowed cash from bank in order to purchase office equipment. Signed a 60-day note payable to the bank, $1,500.
       9 Purchased office equipment, paying cash, $1,500.

10 Collected $5,000 from customers on account.

12 Purchased insurance for the new office equipment, $250.

13 Sold merchandise to customers on account, $7,500.

14 Received a 30-day note from a customer on account, $2,000.

15 Paid taxes owed at end of December, 1974.

20 Cash sales for the period Jan. 5–Jan. 19, $2,550.

22 Paid salaries to employees, $800.

23 Paid to creditors on account, $9,000.

25 Purchased merchandise for cash, $500.

27 Paid miscellaneous expenses, $130.

29 Collected from customers on account, $4,000.

31 Insurance company refunded overpayment of premium, $30.

REQUIRED. Using a general journal, journalize the transactions of the Kinnison Kwik Service Company during January, 1975 (omit posting reference).

3-3. *REQUIRED*

(a) Using the balance-column account form illustrated in the chapter, post to the general ledger the January 1975 transactions of the Kinnison Kwik Service Company from the general journal in Problem 3-2. (Be sure to include opening balances.)

(b) Prepare a trial balance as of January 31, 1975, for the Kinnison Kwik Service Company.

3-4. George Smith, as bookkeeper for the Kinnison Kwik Service Company, heard that it is quicker to use a sales journal, a cash receipts journal, a purchases journal, and a cash payments journal in addition to a general journal.

REQUIRED

(a) To test this hypothesis, journalize the January 1975 transactions in Problem 3-2, using a general journal and the four special journals. Use the account numbers from the chapter.

(b) Post to the general ledger.

3-5. The Ross Company, Inc. completed the following transactions during May 1974:

May 1 Sydney Ross paid $40,000 cash for the common stock of the corporation.

5 Purchased land for $12,000 and a building for $18,000. A down payment of $16,000 was made and a promissory note was signed for the remainder.

9 Purchased office supplies of $450 on account.

13 Purchased office furniture from Bernard Furniture Company for $5,000. One-half of the purchase price was paid in cash with the remainder to be paid in 14 days.

17 Sold one-half of the land purchased on May 5 to Tom Patterson for $6,000, due in six months.

27 Paid the remainder owed to Bernard Furniture Company.

30 Paid the amount owed on the office supplies.

REQUIRED

(a) Journalize the transactions of Ross Company, Inc. for the month of May in a general journal.

(b) Using the balance-column account form illustrated in the chapter, post the May transactions to the general ledger. For posting reference purposes, use your own numbering system to number the accounts.

(c) Prepare a trial balance as of May 31, 1974, to make sure your general ledger is in balance.

3-6. The following transactions were made by Central Distributors, Inc. during February 1974:

Feb. 1 Purchased office supplies for cash, $250.
4 Purchased merchandise on account, $1,400.
5 Purchased office equipment on account, $2,000.
8 Sold merchandise to customers for cash, $850.
11 Sold merchandise to customers on account, $1,185.
13 Paid for merchandise purchased on Feb. 4, $1,400.
15 Paid one-half of amount due for office equipment, $1,000.
17 Purchased a three-year fire insurance policy covering the office equipment, $372.
18 Paid semimonthly salaries, $4,600.
21 Sold merchandise to customers on account, $1,500.
23 Collected from customer on account, $785.
26 Received credit for returning some of the office equipment which was defective, $450.
28 Sold merchandise to customers for cash, $675.
28 Paid miscellaneous expenses, $250.

REQUIRED

(a) Journalize the preceding transactions. Central Distributors, Inc. uses only a general journal.

(b) Post the transactions after setting up ledger accounts with the following balances:

101	Cash	$13,855
110	Accounts receivable	–0–
121	Office supplies	125
131	Merchandise inventory	1,250
141	Prepaid insurance	500
151	Office equipment	9,750
201	Accounts payable	–0–
301	Common stock	20,000
351	Retained earnings	6,380
401	Sales	3,600
531	Salary expense	4,300
591	Miscellaneous expense	200

(c) Prepare a trial balance as of February 28, 1974.

3-7. The Fox Company, which sells office equipment and supplies, completed the following transactions during November 1974:

Nov. 2 Sold merchandise to the Giles Company in the amount of $850 on account, invoice no. 331.

3 Sold merchandise to Jesse Jones in the amount of $145 for cash.

9 Sold merchandise on account to Simmons, Inc. in the amount of $4,350, invoice no. 332.

12 Collected amount due from the Giles Company.

15 Sold merchandise to James Wiley on account for $1,500, invoice no. 333.

18 Collected $4,020 on notes receivable including $20 interest.

21 Miscellaneous cash sales for the day amounted to $333.

22 Collected the amount due from Simmons, Inc.

25 Collected the amount due from James Wiley.

27 Sold merchandise on account to King Kernel Corporation in the amount of $4,000, invoice no. 334.

28 Sold merchandise on account to Mitchell Motors, Inc. in the amount of $750, invoice no. 335.

REQUIRED. Using a sales journal and a cash receipts journal as illustrated in the chapter, journalize the preceding transactions for the month of November 1974.

3-8. The Fox Company, in addition to the transactions in the previous problem, also completed the following transactions during November 1974:

Nov. 1 Purchased merchandise on account from Bonner Desks, Inc. in the amount of $2,250.

5 Purchased merchandise from Scot Company and paid $925 in cash, check no. 1437.

10 Purchased merchandise on account from Handsum Furniture Corporation in the amount of $1,495.

11 Paid amount due to Bonner Desks, Inc., check no. 1438.

16 Paid salary expense of $6,500, check no. 1439.

19 Purchased supplies on account from the Hevvy Paper Company in the amount of $80.

20 Purchased supplies for cash from the Sullivan Supply House in the amount of $135, check no. 1440.

20 Purchased merchandise on account from Bonner Desks, Inc. in the amount of $4,400.

26 Purchased office equipment on account from Equipment Unlimited, Inc. in the amount of $5,000

30 Paid amount due to Bonner Desks, Inc., check no. 1441.

REQUIRED. Using a purchases journal and a cash payments journal as illustrated in the chapter, journalize the preceding transactions for the month of November 1974.

3-9. The Fox Company's bookkeeper is inexperienced. He has journalized the transactions for the month of November 1974 (Problems 3-7 and 3-8). However, he is not sure about the posting procedures. Fox Company asks you to demonstrate the posting procedures to him using the following steps:

REQUIRED

(a) Set up subsidiary Accounts Receivable accounts as follows: Giles Company, King Kernel Corporation, Mitchell Motors, Inc., Simmons, Inc., and James Wiley. None of these accounts have balances as of November 1.

(b) Set up subsidiary Accounts Payable accounts as follows: Bonner Desks, Inc., Equipment Unlimited, Inc., Handsum Furniture Corporation and Hevvy Paper Company. None of these accounts have balances as of November 1.

(c) Set up general ledger accounts with the following balances:

Acct. No.	Acct. Name	Balance
101	Cash	$26,000
110	Accounts receivable	–0–
115	Notes receivable	4,000
121	Supplies	650
131	Merchandise inventory	–0–
151	Office equipment	2,200
201	Accounts payable	–0–
301	Common stock	25,000
351	Retained earnings	7,850
401	Sales	–0–
491	Interest income	–0–
531	Salary expense	–0–

(d) Make the daily postings in chronological order.
   (1) Post to the subsidiary ledgers from the accounts receivable and accounts payable columns of all four journals.
   (2) Post to the general ledger.

(e) To determine if you are still in balance, prepare a trial balance of general ledger accounts. Also determine if the total of each subsidiary ledger is the same as the balance of its respective general ledger account.

Conceptual Problems

3-10. Motes Sporting Goods, Inc. began operations on January 2, 1974. On January 31, a fire destroyed part of the office. The general journal was destroyed in the fire, but the general ledger was saved and appears in Figure 3-9.

REQUIRED

(a) From the information contained in the general ledger, reconstruct the general journal.

(b) Prepare a trial balance as of January 31, 1974.

3-11. The bookkeeper for the Southwest Corporation, Bob Bando, has little formal accounting education. However, he is a very practical man and has

Cash

ACCOUNT NO. 101

DATE	EXPLANATION	POST. REF.	DEBIT	CREDIT	BALANCE
1974					
Jan. 2		GJ1	100,000		100,000
3		GJ1		15,000	85,000
15		GJ1		320	84,680
21		GJ1		350	84,330
22		GJ1	12,000		96,330
22		GJ1		3,400	92,930
25		GJ2		10,000	82,930
28		GJ2	13,000		95,930
30		GJ2		480	95,450

Accounts Receivable

ACCOUNT NO. 110

DATE	EXPLANATION	POST. REF.	DEBIT	CREDIT	BALANCE
1974					
Jan. 12		GJ1	14,300		14,300
22		GJ1		12,000	2,300

Supplies

ACCOUNT NO. 121

DATE	EXPLANATION	POST. REF.	DEBIT	CREDIT	BALANCE
1974					
Jan. 21		GJ1	200		200

Merchandise Inventory

ACCOUNT NO. 131

DATE	EXPLANATION	POST. REF.	DEBIT	CREDIT	BALANCE
1974					
Jan. 10		GJ1	22,000		22,000

Prepaid Insurance

ACCOUNT NO. 141

DATE	EXPLANATION	POST. REF.	DEBIT	CREDIT	BALANCE
1974					
Jan. 30		GJ2	480		480

FIGURE 3-9

5   PROBLEMS

Buildings ACCOUNT NO. 161

DATE	EXPLANATION	POST. REF.	DEBIT	CREDIT	BALANCE
1974 Jan. 3		GJ1	150,000		150,000

Accounts Payable ACCOUNT NO. 201

DATE	EXPLANATION	POST. REF.	DEBIT	CREDIT	BALANCE
1974 Jan. 10		GJ1		22,000	22,000
25		GJ2	10,000		12,000

Mortgage Payable ACCOUNT NO. 241

DATE	EXPLANATION	POST. REF.	DEBIT	CREDIT	BALANCE
1974 Jan. 3		GJ1		135,000	135,000

Common Stock ACCOUNT NO. 301

DATE	EXPLANATION	POST. REF.	DEBIT	CREDIT	BALANCE
1974 Jan. 2		GJ1		100,000	100,000

Sales ACCOUNT NO. 401

DATE	EXPLANATION	POST. REF.	DEBIT	CREDIT	BALANCE
1974 Jan. 12		GJ1		14,300	14,300
28		GJ2		13,000	27,300

Salary Expense ACCOUNT NO. 531

DATE	EXPLANATION	POST. REF.	DEBIT	CREDIT	BALANCE
1974 Jan. 22		GJ1	3,400		3,400

FIGURE 3-9 Continued

Repairs Expense					ACCOUNT NO. 552
DATE	EXPLANATION	POST. REF.	DEBIT	CREDIT	BALANCE
1974 Jan. 15		GJ1	320		320

Miscellaneous Expense					ACCOUNT NO. 591
DATE	EXPLANATION	POST. REF.	DEBIT	CREDIT	BALANCE
1974 Jan. 21		GJ1	150		150

FIGURE 3-9 Continued

learned the existing bookkeeping system through on-the-job experience. The existing system consists of a general journal and a general ledger.

Since 70% of Southwest Corporation's sales are made on account, you have recommended as additions to the system a sales journal, a cash receipts journal, and an accounts receivable subsidiary ledger. Bob Bando disagrees with your suggestions. He says that the recommended system would be too confusing. He is afraid that he will not know in which journal to record a particular entry. He also is concerned that journalizing transactions will take longer since the recording process is different for each journal. With three journals and two ledgers, Bob is convinced that the posting procedure would cause more errors and would require more time than the present system. Finally, he argues that, with an accounts receivable ledger including an account for each customer, it would be almost impossible to prepare a trial balance.

REQUIRED. Counter each of Bob's contentions by pointing out the advantages of special journals and subsidiary ledgers.

3-12. The bookkeeper for Harper Wholesalers, Inc. prepared a trial balance as of December 31, 1974.

An examination of the records reveals the following:

(a) A collection of an account of $350 was erroneously recorded in the cash receipts journal as a debit to Cash and a credit to Accounts Receivable of $530.

(b) A payment of $50 for taxes due to the state was recorded in the general journal as a debit to Taxes Payable and a credit to Cash for $50.

(c) The purchase of land of $1,000 was recorded in the cash payments journal as a debit to Buildings and a credit to Cash for $1,000.

(d) A purchase of office equipment of $1,200 was recorded in the cash receipts journal as a debit to Cash and a credit to Office Equipment for $1,200.

Harper Wholesalers, Inc.
Trial Balance
December 31, 1974

	Debit	Credit
Cash	$19,620	
Accounts receivable	5,470	
Merchandise inventory	14,500	
Prepaid insurance	100	
Office equipment	600	
Buildings	10,000	
Accounts payable		$ 9,390
Salaries payable		200
Taxes payable		150
Common stock		22,400
Retained earnings		14,350
	$50,290	$46,490

(e) A payment on account of $320 was erroneously recorded in the cash payments journal as a debit to Accounts Payable for $230 and a credit to Cash for $230.

(f) The sale of common stock of $4,000 was recorded properly in the cash receipts journal, but the credit to Common Stock was posted to the general ledger as a credit of $400.

(g) The payment of an insurance premium of $100 was recorded properly in the cash payments journal but was posted to the general ledger as a debit to Cash and a debit to Prepaid Insurance for $100.

REQUIRED. For each of the preceding observations, state whether the error caused the trial balance to be out of balance and indicate how the books should be corrected. Also prepare a corrected trial balance.

4

THE ADJUSTMENT
AND CLOSING PROCESS

EDUCATIONAL OBJECTIVES

The material in this chapter is designed to achieve several educational objectives. These include:

1. A basic understanding of why adjusting entries may be necessary at the end of an accounting period.

2. The ability to determine the necessary adjustments and how to make the adjusting entries required.

3. A more complete understanding of the relationship of the income statement and balance sheet through an analysis of the comprehensive illustration of the accounting cycle.

4. A knowledge of the effect of the closing process on the Revenue and Expense accounts and the Retained Earnings account.

In Chapter 3, the emphasis was on recording in the journal and posting to the ledger those transactions which occurred during the accounting period. The purpose of this chapter is to complete the discussion of the accounting cycle; therefore, the emphasis will be on those activities which are performed at the end of the period. In addition, this chapter will include a comprehensive illustration of the accounting cycle.

ADJUSTING THE ACCOUNTS

Throughout the previous chapter, reference was made to the recording and posting of transactions as they occurred. However, accounting is also concerned with the recognition of certain data which are not the result of current transactions. Entries are sometimes required because of the passage of time. These situations are related to some previous transactions in that the entries reflect the effect of the prior transaction on the current period financial statements. In the present context, these situations are referred to as *economic events* to distinguish them from the original transaction.

These economic events must be recorded in order to have a proper matching of revenues and expenses for income statement purposes and to achieve a proper valuation for balance sheet purposes. During the adjusting process, each account is brought up-to-date by the recording of those events which are necessary to achieve a proper matching and valuation. Therefore, prior to the preparation of financial statements, it is necessary to review the general ledger accounts in order to determine if there have been economic events which affect the accounts but have not been recorded. If such events have occurred, it is necessary to determine the accounts to be debited and credited as well as the amount of the entry to be recorded in the journal.

In addition to the events which have occurred, there is the possibility that a transaction was recorded on the basis of certain expectations which failed to materialize. For example, cash may have been received in advance for services to be performed during the period; however, if the service was not performed in the period, an adjustment is necessary. All adjustments have one thing in common: *each adjustment affects one or more balance sheet accounts and one or more income statement accounts.*

The four major types of adjustments affect the following components of the balance sheet and income statement:

1. EXPENSES and ASSETS
2. REVENUES and ASSETS
3. EXPENSES and LIABILITIES
4. REVENUES and LIABILITIES

Expenses and Assets

Three typical adjustments will be presented to illustrate the adjustment process for expenses and assets. They are:

1. Depreciation
2. Insurance
3. Inventory

Depreciation. An adjustment which always will be necessary relates to the plant and equipment of the firm. In this case, the purchase constituted the transaction and would have been recorded as follows, assuming that the building costs $30,000:

Buildings	30,000	
Cash		30,000

The debit entry to Buildings reflects the cost of the new asset. At the end of the accounting period the building is still available for future use, but it does have a limited life. The cost should be allocated over the useful life; otherwise the firm would continue to reflect as an asset the full original cost.

Since the building provides benefits in each of the periods during its life, a portion of the cost should be recorded as an expense, to be matched with the revenue earned during the period. There are several methods available for allocating the cost and they will be discussed in more detail in Chapter 9. For illustrative purposes, assume that the building is expected to be useful for 30 years and that an equal amount is to be allocated to each year. The adjusting journal entry necessary to reflect this economic event is:

Depreciation Expense	1,000	
Accumulated Depreciation —		
Buildings		1,000

When recording depreciation, the usual practice is to credit an account titled Accumulated Depreciation rather than the particular asset. This account is used so that the asset account may continue to reflect the cost of the asset. The historical cost is considered to be useful information because the relationship between the accumulated depreciation account balance and the historical cost informs the user about the percentage of cost allocation. When determining the asset amount to be shown on the balance sheet, accumulated depreciation is deducted from the corresponding asset account. An account which serves to offset another account is referred to as a *contra* account. For example, the Accumulated Depreciation — Buildings account is a *contra* account to the Buildings account. The balance sheet at the end of each year would reflect the following information for the building and the related cost allocation.

Fixed assets

Buildings (historical cost)	xxx
Less: Accumulated depreciation	xx
Net fixed asset	xx

Insurance. A second example of an economic event which must be recorded is the expiration of prepaid insurance. For example, assume that on January 1, 1974 the firm paid $1,200 for a three-year fire insurance policy. At that time, the entry to record the transaction would have been:

Prepaid Insurance	1,200	
Cash		1,200

At the end of the first year, the company still has insurance protection, but it no longer has three year's protection and should not show the original cost as an asset.

In order for the accounts to reflect the appropriate balance, one-third of the premium should be shown as insurance expense for the current period. The adjusting entry to reflect the expired portion of the asset is:

Insurance Expense	400	
Prepaid Insurance		400

The remaining two-thirds, or $800, should continue to be shown as Prepaid Insurance on the balance sheet.

Inventory. A third example of an account which needs to be brought up-to-date is the Inventory. This account will be discussed thoroughly in Chapter 7, but it serves as an illustration of a somewhat different adjustment than insurance and depreciation. At the end of the accounting period, the normal practice is to count the merchandise on hand. Once the inventory has been counted and valued, there will be a difference between the inventory account balance and the inventory actually present, because of accounting procedures.

In all of the illustrations thus far, the purchase of merchandise has been debited to the Merchandise Inventory account. As sales were made, the only entry at that time was a debit to Accounts Receivable and a credit to Sales for the selling price. The Merchandise Inventory account was not reduced by the cost of those sales. Thus, at the end of the period, an adjustment will be required in order to reduce the Merchandise Inventory account and to establish an expense for the cost of goods sold.

To illustrate the adjustment, assume that the Merchandise Inventory account has a balance of $38,250 and that a physical count indicates an actual inventory of $14,400. The implication is that merchandise in the amount of $23,850 was sold during the period. The appropriate adjusting entry would be:

        Cost of Goods Sold   23,850
            Merchandise Inventory   23,850

The Cost of Goods Sold account, as illustrated in Chapter 2, is an expense account and is deducted from sales to determine the gross profit on sales.

In addition to these three major types of adjustment, certain other adjustments may be necessary because some companies follow the practice of initially recording minor cost outlays as an expense. This procedure may occur because of company policy or because it is anticipated that the expenditure will benefit only the current period.

As an example, assume that supplies of $600 were purchased on March 1 and recorded as follows:

        Supplies Expense   600
            Cash   600

If $50 of supplies were found to be on hand at the end of the period, the following adjusting entry would be required in order to establish the asset amount:

        Supplies   50
            Supplies Expense   50

This adjustment brings the accounts to the same ending balance which they would have had if the initial recording had been to the Supplies account. The adjustment in that case would be:

        Supplies Expense   550
            Supplies   550

As indicated, the purpose of adjusting entries is to bring the accounts up-to-date. Thus, irrespective of the initial recording, the adjusting process serves to bring the accounts to their proper balances.

Revenues and Assets

A common adjustment affecting revenues and assets involves income which has been earned through the passage of time, but which has not yet been collected. The recognition of this economic event requires a debit to a Receivable and a credit to a Revenue account. This situation arises whenever a firm has notes receivable from customers, or other interest bearing securities. For example, assume that the firm had some extra cash available and invested it in various short-term U.S. Treasury notes that will mature in January and February of the following year. At the end of

December, the notes are reviewed and the amount of interest earned from the date of issuance to December 31 is determined to be $135. The adjusting entry to reflect this economic event is:

Interest Receivable	135	
Interest Income		135

The collection of the receivable, created by the adjusting entry, will occur at the specified maturity date. The recording of the receivable does not mean that the amount is now due but that the interest has been earned and should be recognized in the accounts as an asset and a revenue.

Expenses and Liabilities

At the end of the accounting period, there may be expenses which have been incurred, but not recorded. These economic facts also require adjusting entries. A rather common situation pertains to employees salaries. In most firms, there is a delay between the end of a pay period and the actual payment of cash to the employees. At the end of the month, or accounting period, an adjusting entry must be made for any salary which has been earned by the employees, but not paid. Otherwise, the expenses and the liabilities for the month would be understated. Assume that, because of weekly pay periods, on June 30 the firm owes $1,275 to employees for the last two days of the month. The adjusting entry to reflect this would be:

Salary Expense	1,275	
Salaries Payable		1,275

When payment is made in July for the entire pay period, only the portion of payment representing July salaries will be recorded as an expense. For example, assume that the salaries for the week ending July 3 are $3,100. The salary payment would be recorded as:

Salaries Payable	1,275	
Salaries Expense	1,825	
Cash		3,100

Other entries of this type may be necessary for interest, utility bills, income taxes, and rent.

Revenues and Liabilities

When cash is received prior to being earned the firm has an obligation, or liability, to perform some service in the future. This situation is evident in the case of a magazine publishing company where the common practice is the receipt of payment in advance.

As an example of this situation, assume that the Ace Publishing Company received $24 on January 1, for a two-year subscription to its magazine. The receipt of cash would be recorded as follows:

Cash	24	
Unearned Subscriptions		24

If financial statements are prepared at the end of the year, the following adjustment is necessary:

Unearned Subscriptions	12	
Subscription Revenue		12

At the time the cash was received, Ace could have recorded the transaction as:

Cash	24	
Subscription Revenue		24

If this entry was made, the adjusting entry would be:

Subscription Revenue	12	
Unearned Subscription		12

In either case the subscription revenue for the year is $12 and the balance in the liability, Unearned Subscription, at the end of the first year is $12.

Two common terms that are used in accounting terminology relative to the adjusting process are *accrued* and *deferred.* Accrued refers to a revenue which has been earned but not received or an expense which has been incurred but not paid. For example, unpaid salaries and interest earned are items to be accrued. Deferred refers to revenue which has been received but not earned and prepaid expenses. For example, payments received in advance such as magazine subscriptions and prepayments such as insurance are items to be deferred.

Summary of Adjustments

There are several other adjustments which may be necessary at the end of the accounting period. Certain adjustments are always a part of the accounting cycle while others occur periodically. The basic rule to follow relative to adjustments is: *review each of the accounts in the general ledger and determine whether the account has the appropriate balance. If not, the necessary adjustment must be determined, and recorded, in order to bring the account up-to-date.*

TRANSFER OF TEMPORARY ACCOUNT BALANCES

In the previous chapter, Revenue and Expense accounts were presented as *temporary capital accounts*. These accounts were considered a part of the capital, because all items recorded in a Revenue or Expense account could have been recorded directly as an increase, or decrease, in capital. Several accounts were used in order to accumulate details about revenues and expenses for the preparation of the income statement.

At the end of the accounting period, the Revenue and Expense accounts are summarized and transferred as a net amount to Retained Earnings. This transfer procedure is referred to as *closing the accounts* and is accomplished through closing journal entries.

The closing process may be illustrated by assuming the following account balances:

SALES		SALES RETURNS AND ALLOWANCES	
	102,725	2,105	

COST OF GOODS SOLD		SALARIES EXPENSE	
66,250		26,735	

DEPRECIATION EXPENSE		UTILITIES EXPENSE	
2,340		570	

Each Revenue and Expense account is analyzed to determine the account balance. A closing journal entry is then prepared debiting each account which has a credit balance and crediting each account with a debit balance. Thus, the revenues will be debited, the expenses credited, and the difference entered in the Retained Earnings account. The closing journal entry for the above accounts would be:

Sales	102,725	
Sales Returns and Allowances		2,105
Cost of Goods Sold		66,250
Salaries Expense		26,735
Depreciation Expense		2,340
Utilities Expense		570
Retained Earnings		4,725

To illustrate the effect of the closing process on the temporary capital accounts and the capital account, the account balances and closing entry will be shown in T accounts:

	SALES			SALES RETURNS AND ALLOWANCES	
Closing 102,725	102,725 Bal.		Bal. 2,105	2,105 Closing

	COST OF GOODS SOLD			SALARIES EXPENSE	
Bal. 66,250	66,250 Closing		Bal. 26,735	26,735 Closing

	DEPRECIATION EXPENSE			UTILITIES EXPENSE	
Bal. 2,340	2,340 Closing		Bal. 570	570 Closing

RETAINED EARNINGS	
	−0− Bal.
	4,725 Net Income

The difference between the Revenue and Expense accounts, which is transferred to Retained Earnings, is the net income for the period. If the result of operations for the period is a loss, the entry to Retained Earnings would be a debit, or decrease, since expenses for the period would be greater than the revenue.

The closing process sometimes involves an additional step which was not taken in the preceding illustration. This step involves the use of an income summary to accumulate the revenue and expense account balances. Then the balance in the Income Summary account, which is the difference between the revenues and expenses, is transferred to the Retained Earnings account.

Continuing with the preceding example, the closing journal entry utilizing an Income Summary account would be as follows:

Sales	102,725	
Sales Returns and Allowances		2,105
Cost of Goods Sold		66,250
Salaries Expense		26,735
Depreciation Expense		2,340
Utilities Expense		570
Income Summary		4,725

The balance in the Income Summary account is then transferred to Retained Earnings as follows:

Income Summary	4,725	
Retained Earnings		4,725

Thus the closing process serves to reduce the accounting equation from:

$$A = L + [C + (R - E)] \quad \text{to} \quad A = L + C$$

The detailed procedural aspects of the accounting cycle are completed when the net balance of the temporary capital accounts have been transferred to Retained Earnings. The entire accounting cycle, including the preparation of financial statements, will now be illustrated.

COMPREHENSIVE ILLUSTRATION OF THE ACCOUNTING CYCLE

In this illustration, the Random Corporation will be used along with a revised version of the chart of accounts from Chapter 3.

Random Corporation
Chart of Accounts

100 Assets
    101   Cash
    110   Accounts receivable
    111   Allowance for uncollectible accounts
    121   Supplies
    131   Merchandise inventory
    141   Prepaid insurance
    161   Buildings
    171   Accumulated depreciation — buildings

200 Liabilities
    201   Accounts payable
    212   Federal income taxes withheld
    213   Federal social security taxes withheld
    241   Notes payable
    261   Utilities payable

300 Capital
    301   Common stock
    351   Retained earnings

400 Revenue
    401   Sales
    451   Sales returns and allowances

500 Expenses
    501   Cost of goods sold
    531   Salaries expense
    541   Supplies expense
    545   Utilities expense
    551   Depreciation expense
    571   Insurance expense
    581   Bad debt expense
    591   Miscellaneous expense

In order to minimize the number of transactions to be recorded, the illustration will include only the month of June, 1974. Assume that financial statements are prepared, and the temporary capital accounts closed, at the end of each month. Therefore, as a starting point the trial balance of the Random Corporation as of May 31, 1974 is presented:

Random Corporation
Trial Balance
May 31, 1974

	ACCOUNT	DEBITS	CREDITS
101	Cash	$ 9,850	
110	Accounts receivable	15,000	
111	Allowance for uncollectible accounts		$ 150
121	Supplies	300	
131	Merchandise inventory	13,000	
141	Prepaid insurance	500	
161	Buildings	50,000	
171	Accumulated depreciation — buildings		5,000
201	Accounts payable		8,700
241	Notes payable		8,000
301	Common stock		60,000
351	Retained earnings		6,800
		$88,650	$88,650

All of the accounts reflected in the trial balance as of May 31 are shown with a beginning balance in the illustration of the general ledger. When transactions are recorded in the journal, an explanation may be given following each complete entry. In this illustration, sufficient explanation is given in the transaction and it is not repeated in the journal.

Transactions

The following transactions of the Random Corporation which occurred during June 1974 will be recorded in the journal and posted to the appropriate general ledger accounts:

June 3 Purchased supplies in the amount of $475 from City Supply Company. The purchase was made on credit; therefore, the transaction results in an increase in the asset, Supplies, and an increase in the liability, Accounts Payable.

June 4 Paid to various creditors $3,000 of the total accounts payable outstanding at the beginning of the month. (In practice, the individual creditors would be identified since the amount of payment made to each must be posted to the individual account.) This transaction decreases the asset, Cash, and also decreases the liability, Accounts Payable.

June 6 Purchased on credit, merchandise in the amount of $18,300 from the Leesburg Company. This transaction is an increase in the asset, Merchandise Inventory, and an increase in the liability, Accounts Payable.

June 7 Total sales for the week were $23,600. The cash sales were $4,700 and the credit sales were $18,900. (In this example, the individual customer names are not provided; therefore, the subsidiary accounts cannot be posted.) The cash sales result in an increase in the Cash account while the credit sales result in an increase in the Accounts Receivable account. The total sales represent an increase in capital but are recorded in the temporary revenue account, Sales.

June 7 Received and paid an invoice for $375 from Allied Freight Company. The freight costs pertained to the merchandise purchased on June 6. This transaction results in an increase in the Merchandise Inventory and a decrease in Cash.

June 10 Paid miscellaneous expenses in the amount of $350. (Most firms have an account for items which occur so infrequently, or which are so small, that separate expense accounts are not warranted.) The transaction is recorded as an increase in Miscellaneous Expense and a decrease in Cash.

June 12 Collections from customers as payment on accounts receivable during the past several days were $16,800. The payments are recorded as a reduction of Accounts Receivable and an increase in Cash. In addition, the individual customer's accounts are reduced by the amount of their payment.

June 14 Salaries for the first half of the month in the amount of $8,900 were paid. Federal income taxes and social security taxes of $1,240 and $405 were withheld from the employees in accordance with tax regulations. These withheld amounts are liabilities of the firm which must be paid to the federal government. The transaction is recorded as an increase in Salaries Expense and the tax liabilities and a decrease in Cash for the actual amount paid to the employees.

June 14 Total sales for the week were $22,800. The cash sales were $4,400.

June 17 Received and paid an invoice for utilities. The amount was $530 and is recorded as an increase in Utilities Expense and a decrease in Cash.

June 17 Purchased on credit, merchandise in the amount of $17,500 from the White Company.

June 19 Collected $15,700 from customers as payments on accounts receivable.

June 21 Returned merchandise with an invoice price of $825 to the White Company. This transaction is recorded as a decrease in Merchandise Inventory and a decrease in Accounts Payable.

June 21 Total sales for the week were $24,100 and credit sales were $19,650.

June 24 Paid the City Supply Company for supplies purchased on June 3, in the amount of $475.

June 25 Purchased on credit, merchandise in the amount of $27,000 from the Leesburg Company.

June 26 Paid the Leesburg Company $18,300 for the merchandise purchased June 6 on credit.

June 27 Collected $19,500 from customers on accounts receivable.

June 27 Received defective merchandise from customer. The merchandise was purchased on credit and had an invoice price of $980. This transaction is recorded as an increase in the *contra* account Sales Returns and Allowances and a decrease in Accounts Receivable.

June 27 Paid White Company $17,500 for merchandise purchased June 17 on credit.

June 28 Total sales for the week were $28,250 and cash sales were $6,475.

June 28 Salaries for the second half of the month in the amount of $9,200 were paid. Federal income taxes of $1,280 and social security taxes of $430 were withheld.

Adjusting information as of June 30

1. Merchandise inventory on hand, as determined by a physical count, was $15,500. The account balance must be decreased, since the account balance, per books, is $75,350. The difference, $75,350 − $15,500 = $59,850, must be transferred to the Cost of Goods Sold account.

2. Bad debts for the year are estimated to be 1/2 of 1% of net sales. The total sales for the month of $98,750 less the returns of $980, multiplied by 0.5%, give the amount of the adjustment. The entry is an increase in Bad Debt Expense and an increase in the Allowance for Uncollectible Accounts.

3. The building is estimated to have a useful life of 25 years. The total cost of $50,000 divided by 25 years results in a charge to the current month of $167 for Depreciation Expense and an increase in the Accumulated Depreciation − Buildings.

	DATE 1974		DESCRIPTION	POST. REF.	DEBIT	CREDIT	
1	June	3	Supplies	121	475		1
2			Accounts Payable	201		475	2
3							3
4		4	Accounts Payable	201	3000		4
5			Cash	101		3000	5
6							6
7		6	Merchandise Inventory	131	18300		7
8			Accounts Payable	201		18300	8
9							9
10		7	Accounts Receivable	110	18900		10
11			Cash	101	4700		11
12			Sales	401		23600	12
13							13
14		7	Merchandise Inventory	131	375		14
15			Cash	101		375	15
16							16
17		10	Miscellaneous Expense	591	350		17
18			Cash	101		350	18
19							19
20		12	Cash	101	16800		20
21			Accounts Receivable	110		16800	21
22							22
23		14	Salaries Expense	531	8900		23
24			Cash	101		7255	24
25			Federal Income Taxes	212		1240	25
26			Withheld				26
27			Federal Social Security	213		405	27
28			Taxes Withheld				28
29							29
30		14	Accounts Receivable	110	18400		30
31			Cash	101	4400		31
32			Sales	401		22800	32
33							33

	DATE 1974		DESCRIPTION	POST. REF.	DEBIT	CREDIT	
1	June	17	Utilities Expense	545	530		1
2			Cash	101		530	2
3							3
4		17	Merchandise Inventory	131	17500		4
5			Accounts Payable	201		17500	5
6							6
7		19	Cash	101	15700		7
8			Accounts Receivable	110		15700	8
9							9
10		21	Accounts Payable	201	825		10
11			Merchandise Inventory	131		825	11
12							12
13		21	Accounts Receivable	110	19650		13
14			Cash	101	4450		14
15			Sales	401		24100	15
16							16
17		24	Accounts Payable	201	475		17
18			Cash	101		475	18
19							19
20		25	Merchandise Inventory	131	27000		20
21			Accounts Payable	201		27000	21
22							22
23		26	Accounts Payable	201	18300		23
24			Cash	101		18300	24
25							25
26		27	Cash	101	19500		26
27			Accounts Receivable	110		19500	27
28							28
29		27	Sales Returns and Allowances	451	980		29
30			Accounts Receivable	110		980	30
31							31
32							32
33							33

	DATE 1974		DESCRIPTION	POST. REF.	DEBIT	CREDIT	
1	June	27	Accounts Payable	201	17500		1
2			Cash	101		17500	2
3							3
4		28	Accounts Receivable	110	21775		4
5			Cash	101	6475		5
6			Sales	401		28250	6
7							7
8		28	Salaries Expense	531	9200		8
9			Cash	101		7490	9
10			Federal Income Taxes Withheld	212		1280	10
11			Federal Social Security				11
12			Taxes Withheld	213		430	12
13							13
14		30	Cost of Goods Sold	501	59850		14
15			Merchandise Inventory	131		59850	15
16			ADJUSTING ENTRIES				16
17		30	Bad Debt Expense	581	489		17
18			Allowance for				18
19			Uncollectible Accounts	111		489	19
20							20
21		30	Depreciation Expense	551	167		21
22			Accumulated Depre-				22
23			ciation — Buildings	171		167	23
24							24
25		30	Insurance Expense	571	100		25
26			Prepaid Insurance	141		100	26
27							27
28		30	Supplies Expense	541	425		28
29			Supplies	121		425	29
30							30
31		30	Utilities Expense	545	125		31
32			Utilities Payable	261		125	32
33							33

	DATE 1974		ITEM	POST. REF.	DEBIT	CREDIT	BALANCE DEBIT	BALANCE CREDIT	
1	June	1	Balance	✔			9850		1
2		4		GJ96		3000	6850		2
3		7		GJ96	4700		11550		3
4		7		GJ96		375	11175		4
5		10		GJ96		350	10825		5
6		12		GJ96	16800		27625		6
7		14		GJ96		7255	20370		7
8		14		GJ96	4400		24770		8
9		17		GJ97		530	24240		9
10		19		GJ97	15700		39940		10
11		21		GJ97	4450		44390		11
12		24		GJ97		475	43915		12
13		26		GJ97		18300	25615		13
14		27		GJ97	19500		45115		14
15		27		GJ98		17500	27615		15
16		28		GJ98	6475		34090		16
17		28		GJ98		7490	26600		17
18									18

	DATE 1974		ITEM	POST. REF.	DEBIT	CREDIT	BALANCE DEBIT	BALANCE CREDIT	
1	June	1	Balance	✔			15000		1
2		7		GJ96	18900		33900		2
3		12		GJ96		16800	17100		3
4		14		GJ96	18400		35500		4
5		19		GJ97		15700	19800		5
6		21		GJ97	19650		39450		6
7		27		GJ97		19500	19950		7
8		27		GJ97		980	18970		8
9		28		GJ98	21775		40745		9
10									10

ACCOUNT Allowance for Uncollectible Accounts   ACCOUNT NO. 111

	DATE 1974		ITEM	POST. REF.	DEBIT	CREDIT	BALANCE DEBIT	BALANCE CREDIT	
1	June	1	Balance	✓				150	1
2		30	Adjusting entry	GJ98		489		639	2
3									3

ACCOUNT Supplies   ACCOUNT NO. 121

	DATE 1974		ITEM	POST. REF.	DEBIT	CREDIT	BALANCE DEBIT	BALANCE CREDIT	
1	June	1	Balance	✓			300		1
2		3		GJ96	475		775		2
3		30	Adjusting entry	GJ98		425	350		3
4									4

ACCOUNT Merchandise Inventory   ACCOUNT NO. 131

	DATE 1974		ITEM	POST. REF.	DEBIT	CREDIT	BALANCE DEBIT	BALANCE CREDIT	
1	June	1	Balance	✓			13000		1
2		6		GJ96	18300		31300		2
3		7		GJ96	375		31675		3
4		17		GJ97	17500		49175		4
5		21		GJ97		825	48350		5
6		25		GJ97	27000		75350		6
7		30	Adjusting entry	GJ98		59850	15500		7
8									8

ACCOUNT Prepaid Insurance   ACCOUNT NO. 141

	DATE 1974		ITEM	POST. REF.	DEBIT	CREDIT	BALANCE DEBIT	BALANCE CREDIT	
1	June	1	Balance	✓			500		1
2		30	Adjusting entry	GJ98		100	400		2
3									3

ACCOUNT Buildings   ACCOUNT NO. 161

	DATE 1974		ITEM	POST. REF.	DEBIT	CREDIT	BALANCE DEBIT	BALANCE CREDIT	
1	June	1	Balance	✓			50000		1
2									2

ACCOUNT Accumulated Depreciation — Buildings   ACCOUNT NO. 171

	DATE 1974		ITEM	POST. REF.	DEBIT	CREDIT	BALANCE DEBIT	BALANCE CREDIT	
1	June	1	Balance	✓				5000	1
2		30	Adjusting Entry	GJ98		167		5167	2
3									3

ACCOUNT Accounts Payable   ACCOUNT NO. 201

	DATE 1974		ITEM	POST. REF.	DEBIT	CREDIT	BALANCE DEBIT	BALANCE CREDIT	
1	June	1	Balance	✓				8700	1
2		3		GJ96		475		9175	2
3		4		GJ96	3000			6175	3
4		6		GJ96		18300		24475	4
5		17		GJ97		17500		41975	5
6		21		GJ97	825			41150	6
7		24		GJ97	475			40675	7
8		25		GJ97		27000		67675	8
9		26		GJ97	18300			49375	9
10		27		GJ98	17500			31875	10
11									11

ACCOUNT Federal Income Taxes Withheld   ACCOUNT NO. 212

	DATE 1974		ITEM	POST. REF.	DEBIT	CREDIT	BALANCE DEBIT	BALANCE CREDIT	
1	June	14		GJ96		1240		1240	1
2		28		GJ98		1280		2520	2
3									3

ACCOUNT Federal Social Security Taxes Withheld   **ACCOUNT NO.** 213

	DATE 1974		ITEM	POST. REF.	DEBIT	CREDIT	BALANCE DEBIT	BALANCE CREDIT	
1	June	14		GJ96		405		405	1
2		28		GJ98		430		835	2
3									3

ACCOUNT Notes Payable   **ACCOUNT NO.** 241

	DATE 1974		ITEM	POST. REF.	DEBIT	CREDIT	BALANCE DEBIT	BALANCE CREDIT	
1	June	1	Balance	✔				8000	1
2									2

ACCOUNT Utilities Payable   **ACCOUNT NO.** 261

	DATE 1974		ITEM	POST. REF.	DEBIT	CREDIT	BALANCE DEBIT	BALANCE CREDIT	
1	June	30	Adjusting entry	GJ98		125		125	1
2									2

ACCOUNT Common Stock   **ACCOUNT NO.** 301

	DATE 1974		ITEM	POST. REF.	DEBIT	CREDIT	BALANCE DEBIT	BALANCE CREDIT	
1	June	1	Balance	✔				60000	1
2									2

ACCOUNT Retained Earnings   **ACCOUNT NO.** 351

	DATE 1974		ITEM	POST. REF.	DEBIT	CREDIT	BALANCE DEBIT	BALANCE CREDIT	
1	June	1	Balance	✔				6800	1
2									2

ACCOUNT Sales   **ACCOUNT NO.** 401

	DATE 1974		ITEM	POST. REF.	DEBIT	CREDIT	BALANCE DEBIT	BALANCE CREDIT	
1	June	7		GJ96		23600		23600	1
2		14		GJ96		22800		46400	2
3		21		GJ97		24100		70500	3
4		28		GJ98		28250		98750	4
5									5

ACCOUNT Sales Returns and Allowances   **ACCOUNT NO.** 451

	DATE 1974		ITEM	POST. REF.	DEBIT	CREDIT	BALANCE DEBIT	BALANCE CREDIT	
1	June	27		GJ97	980		980		1
2									2

ACCOUNT Cost of Goods Sold   **ACCOUNT NO.** 501

	DATE 1974		ITEM	POST. REF.	DEBIT	CREDIT	BALANCE DEBIT	BALANCE CREDIT	
1	June	30	Adjusting entry	GJ98	59850		59850		1
2									2

ACCOUNT Salaries Expense   **ACCOUNT NO.** 531

	DATE 1974		ITEM	POST. REF.	DEBIT	CREDIT	BALANCE DEBIT	BALANCE CREDIT	
1	June	14		GJ96	8900		8900		1
2		28		GJ98	9200		18100		2
3									3

ACCOUNT Supplies Expense   **ACCOUNT NO.** 541

	DATE 1974		ITEM	POST. REF.	DEBIT	CREDIT	BALANCE DEBIT	BALANCE CREDIT	
1	June	30	Adjusting entry	GJ98	425		425		1
2									2

ACCOUNT   Utilities Expense   ACCOUNT NO. 545

	DATE 1974	ITEM	POST. REF.	DEBIT	CREDIT	BALANCE DEBIT	CREDIT	
1	June 17		GJ97	530		530		1
2	30	Adjusting entry	GJ98	125		655		2
3								3

ACCOUNT   Depreciation Expense   ACCOUNT NO. 551

	DATE 1974	ITEM	POST. REF.	DEBIT	CREDIT	BALANCE DEBIT	CREDIT	
1	June 30	Adjusting entry	GJ98	167		167		1
2								2

ACCOUNT   Insurance Expense   ACCOUNT NO. 571

	DATE 1974	ITEM	POST. REF.	DEBIT	CREDIT	BALANCE DEBIT	CREDIT	
1	June 30	Adjusting entry	GJ98	100		100		1
2								2

ACCOUNT   Bad Debt Expense   ACCOUNT NO. 581

	DATE 1974	ITEM	POST. REF.	DEBIT	CREDIT	BALANCE DEBIT	CREDIT	
1	June 30	Adjusting entry	GJ98	489		489		1
2								2

ACCOUNT   Miscellaneous Expense   ACCOUNT NO. 591

	DATE 1974	ITEM	POST. REF.	DEBIT	CREDIT	BALANCE DEBIT	CREDIT	
1	June 10		GJ96	350		350		1
2								2

4. A review of the schedule of insurance policies indicates unexpired insurance on June 30 to be $400. Consequently, the Prepaid Insurance account should be decreased by $100 and Insurance Expense increased.

5. Supplies on hand determined by physical count were $350. The current balance in the Supplies account must be determined before

an adjustment can be made. In this case, an adjustment of $425 is necessary.

6. A utility bill, for the month of June, in the amount of $125 was received but not paid.

Trial Balance. A trial balance should be prepared before the adjusting entries are recorded in order to insure that the general ledger has equal debits and credits. In addition, another trial balance, an adjusted trial balance, is prepared after the adjusting entries. In this illustration only the adjusted trial balance is presented. The following adjusted trial balance of the Random Corporation gives assurance that debits equal credits *and* provides a convenient source of information for the preparation of the financial statements.

Random Corporation
Adjusted Trial Balance
June 30, 1974

	Account	Debits	Credits
101	Cash	$ 26,600	
110	Accounts receivable	40,745	
111	Allowance for uncollectible accounts		$ 639
121	Supplies	350	
131	Merchandise inventory	15,500	
141	Prepaid insurance	400	
161	Buildings	50,000	
171	Accumulated depreciation — buildings		5,167
201	Accounts payable		31,875
212	Federal income taxes withheld		2,520
213	Federal social security taxes withheld		835
241	Notes payable		8,000
261	Utilities payable		125
301	Common stock		60,000
351	Retained earnings		6,800
401	Sales		98,750
451	Sales returns and allowances	980	
501	Cost of goods sold	59,850	
531	Salaries expense	18,100	
541	Supplies expense	425	
545	Utilities expense	655	
551	Depreciation expense	167	
571	Insurance expense	100	
581	Bad Debt expense	489	
591	Miscellaneous expense	350	
		$214,711	$214,711

Financial Statements

Random Corporation
Income Statement
for the Month of June, 1974

Revenue from sales:		
Sales		$98,750
Less: Sales returns and allowances		980
Net sales		$97,770
Cost of goods sold		59,850
Gross profit on sales		$37,920
Operating expenses:		
Salaries expense	$18,100	
Supplies expense	425	
Utilities expense	655	
Depreciation expense	167	
Insurance expense	100	
Bad debt expense	489	
Miscellaneous expense	350	
Total operating expenses		20,286
Net income from operations		$17,634[a]

Random Corporation
Statement of Retained Earnings
Month of June, 1974

Retained earnings, June 1, 1974	$ 6,800
Net income from operations for the month of June, 1974	17,634
Retained earnings, June 30, 1974	$24,434

[a]For purposes of simplicity, federal and state income taxes and the earnings per share computations are not included.

Random Corporation
Balance Sheet
June 30, 1974

ASSETS

Current Assets

Cash		$26,600
Accounts receivable	$40,745	
Less: Allowance for uncollectible accounts	639	40,106
Supplies		350
Merchandise inventory		15,500
Prepaid insurance		400
Total Current Assets		$82,956

Plant Assets

Buildings	$50,000	
Less: Accumulated depreciation — buildings	5,167	
Total Plant Assets		44,833
TOTAL ASSETS		$127,789

LIABILITIES

Current Liabilities

Accounts payable	$31,875	
Federal income taxes withheld	2,520	
Federal social security taxes withheld	835	
Utilities payable	125	
Total Current Liabilities		$35,355

Long-Term Liabilities

Notes payable		8,000
Total Liabilities		$ 43,355

CAPITAL

Common stock (60,000 shares)	$60,000	
Retained earnings	24,434	
Total Capital		84,434
TOTAL LIABILITIES and CAPITAL		$127,789

COMPREHENSIVE ILLUSTRATION OF THE ACCOUNTING CYCLE

CLOSING THE TEMPORARY ACCOUNTS

The final phase of the accounting cycle is the transfer of the net balance of all temporary capital accounts into Retained Earnings. The formal closing of the temporary accounts of Random Corporation may be accomplished by the following entry:

	DATE 1974		DESCRIPTION	POST. REF.	DEBIT	CREDIT	
1	June	30	Sales		98750		1
2			Sales Returns and Allowances			980	2
3			Salaries Expense			18100	3
4			Supplies Expense			425	4
5			Utilities Expense			655	5
6			Depreciation Expense			167	6
7			Insurance Expense			100	7
8			Bad Debt Expense			489	8
9			Miscellaneous Expense			350	9
10			Cost of Goods Sold			59850	10
11			Retained Earnings			17634	11
12							12

GENERAL JOURNAL — PAGE 99

In the illustration, the closing entry is not posted to the individual ledger accounts.

SUMMARY

In this and the preceding chapter, the fundamentals of accounting have been presented in terms of the conventions and techniques of the accounting cycle. The steps in the accounting cycle can be related to the accounting processes of identification, measurement, and communication of economic information.

Steps in accounting cycle	Process involved
1. Business transactions are analyzed	Identification
2. Transactions are recorded in in the journal	Identification of the nature and measurement of the amount
3. Journal entries are posted to the ledger accounts	Summarization for purposes of communication

Steps in accounting cycle	Process involved
4. Trial balance is prepared	Summarization for purposes of communication
5. Accounts are reviewed and the necessary adjustments determined and recorded in the general journal	Identification and measurement
6. Adjusting journal entries are posted to the ledger	Summarization for purposes of communication
7. Financial statements are prepared from the adjusted accounts	Communication
8. Temporary accounts are closed	Summarization so that the cycle may be continued

The next seven chapters will provide more thorough explanations of each major component of the balance sheet as affected by generally accepted accounting principles.

QUESTIONS

4-1. What is the purpose of adjusting entries?

4-2. What type of accounts are affected by adjusting entries? List the four major types of adjustments based upon the components of each.

4-3. (a) What is the concept of fixed asset depreciation?
(b) Why is the account Accumulated Depreciation credited in the adjusting entry for depreciation rather than the asset account itself?

4-4. On July 1, 1974, ABC Corporation purchased a three-year fire insurance policy on its plant. The entry for the purchase included a debit to Prepaid Insurance for $3,000. If ABC Corporation closes its books on December 31, what adjusting entry should be made for insurance:
(a) At the end of 1974?
(b) At the end of 1975?

4-5. The accountant for the Heidi Company purchased supplies on January 2 every year. He knew the approximate amount of supplies that would be consumed each year, so when the entry was made for the purchase he debited Supplies Expense and credited Cash. In 1975 the plant was closed for two months during which time no supplies were used. The accountant now asks you whether or not an adjusting entry is necessary for the $500 of supplies which remain and, if so, what entry should be made?

4-6. What is the effect on the balance sheet and income statement of a failure to record salaries owed to employees but not paid at year end? Also indicate whether the item in error will be overstated or understated.

4-7. On March 1, Able Company rents a warehouse for one year by paying the $1,200 yearly rent in advance.

(a) Does the transaction increase the assets or expenses of Able Company?

(b) If Able Company operates on a December 31 year end, will this transaction require an adjusting entry at that time?

(c) Is there any justification for debiting an expense account on March 1?

4-8. Given the following information, what accounts would be debited and credited in an adjusting entry?

(a) Salaries are to be recorded as expenses in the year earned.

(b) Inventory at year end should reflect the physical count.

(c) Any interest due should be recorded in the year earned.

(d) Insurance is recorded as an asset when purchased.

(e) Subscriptions from customers are recorded as unearned when received.

(f) Bad debts are estimated at ½ of 1% of net sales.

4-9. (a) Why is it necessary to close the temporary accounts?

(b) Briefly describe how this process is accomplished.

4-10. Which of the following accounts are included in the closing process?

(a) Sales
(b) Accounts receivable
(c) Supplies expense
(d) Rent expense
(e) Cash
(f) Prepaid insurance
(g) Accumulated depreciation
(h) Office equipment
(i) Sales returns and allowances
(j) Interest income

4-11. (a) What is the purpose of an adjusted trial balance?

(b) If the adjusted trial balance shows an equal amount of debits and credits, does this prove that there were no errors in the recording of transactions? Explain.

4-12. What is the order in which financial statements should be prepared? Explain why financial statements must be prepared in this order.

PROBLEMS

Procedural Problems

4-1. The following data relate to the accounts of the Kenneth Corporation:

(a) The Prepaid Insurance account has a $900 balance, $300 of which has expired during the year.

(b) The Office Supplies account has a balance of $1,800. A count of the supplies on hand at year end revealed $400 of supplies still available.

(c) Kenneth Company shows $450 of subscription income in its general ledger. Of this amount, $200 applies to a subsequent period.

(d) Unpaid salaries earned by employees as of the year end amount to $375.

(e) As of the year end, Kenneth Corporation owes $550 of interest on a note which is due in the next accounting period.

REQUIRED. Prepare the adjusting entries necessary as of December 31, the corporation's year end.

4-2. The following data were taken from the records of the Regal Knight Company:

Trial Balance
December 31, 1974

Accounts receivable	$27,500
Allowance for uncollectible accounts	2,000
Accumulated depreciation — buildings	6,500
Accounts payable	15,530
Buildings	36,800
Cash	36,250
Common stock	50,000
Prepaid insurance	1,250
Supplies	3,470
Interest income	850
Rent expense	1,800
Salaries expense	32,000
Sales	87,250
Merchandise inventory	51,150
Retained earnings	28,090

Adjusted Trial Balance
December 31, 1974

Accounts receivable	$27,500
Allowance for uncollectible accounts	2,800
Accumulated depreciation — buildings	7,800
Accounts payable	15,530
Buildings	36,800
Cash	36,250
Common stock	50,000
Prepaid insurance	1,000
Supplies	2,850
Interest income	1,250
Rent expense	1,200
Salaries expense	34,620
Sales	87,250
Merchandise inventory	20,650
Cost of goods sold	30,500
Retained earnings	28,090
Bad Debts expense	800
Depreciation expense	1,300
Insurance expense	250
Supplies expense	620
Interest receivable	400
Prepaid rent	600
Salaries payable	2,620

REQUIRED
(a) Based upon the two trial balances just shown, reconstruct the adjusting entries made by Regal Knight Company on December 31, 1974.

(b) Prepare closing entries for the company.

4-3. *REQUIRED.* Using the information from Problem 4-2, prepare:

(a) Income statement (omit federal and state income taxes).

(b) Balance sheet.

4-4. Following are the trial balance of the Slice and Hook Golf Course at December 31, 1974, the end of their fiscal year, and the necessary adjustment data:

Slice and Hook Golf Course
Trial Balance
December 31, 1974

	Debit	Credit
Cash	$ 8,150	
Golf supplies	4,250	
Prepaid insurance	1,500	
Golf equipment	25,500	
Accumulated depreciation		$ 2,325
Accounts payable		4,600
Common stock		15,000
Retained earnings		6,500
Equipment rental revenue		6,785
Golf fee revenue		21,500
Wages expense	12,500	
Utilities expense	3,840	
Miscellaneous expense	970	
	$56,710	$56,710

Adjustment Data:
- Golf supplies on hand at December 31 amounted to $1,540.
- Prepaid insurance represents a three-year policy purchased on July 1, 1974.
- Depreciation on the golf equipment amounted to $775.
- Included in the Golf Fee Revenue account is $5,000, which represents payments to the golf course for season tickets. The tickets are purchased on April 1 of each year and are valid for a one-year period.
- Wages due but unpaid at year end amount to $300.

REQUIRED
(a) Prepare adjusting entries for the Golf Course as of December 31.

(b) Prepare an income statement, statement of retained earnings, and a balance sheet for 1974.

(c) Close the temporary accounts.

4-5. The following data pertain to the accounts of the Lexington Metal Company for the month of June:

1. The annual insurance premium of $456 was paid on March 1 and the Prepaid Insurance account was debited.
2. Estimated federal and state tax expense for June amounted to $300.
3. The daily payroll is $350, based upon a five-day work week, with employees receiving their pay each Friday. June 30 of this year falls on a Wednesday.
4. The Company holds four $1,000 corporate bonds of Hardin, Inc. that pay 6% interest annually.
5. A customer ordered metal parts which will be manufactured and delivered in July. A deposit of $2,500, received with the order, was credited to sales.
6. A machine, which broke down during June, was repaired at a cost of $550. The bill has not been received.
7. Work has been completed on a special order and the goods have been shipped. The bill, amounting to $3,500, will not be sent until July 1.

REQUIRED
(a) Prepare the adjusting entries to be made by Lexington Metal Company on June 30.
(b) Indicate the effect on Lexington Company's net income if the preceding adjustments are not recorded.

Conceptual Problems

4-6. Each of the following independent cases pertains to the accounts of the Mac Company. Answer the question following each case:
(a) The Office Supplies account showed a balance of $1,800 and $1,450 on December 31, 1973 and 1974, respectively. If $2,500 of office supplies were purchased during 1974, what was the amount of office supplies expense for 1974?
(b) Mac Company has four accounting clerks, each of whom earn a weekly salary of $150. They are paid on Friday of each week, the last day of the five-day work week. What adjusting entry would be made assuming the year ended on a Thursday?
(c) The balance sheets for 1973 and 1974 showed the following balances for prepaid insurance:

> December 31, 1973 $985
> December 31, 1974 $550

The income statement for 1974 showed insurance expense of $1,350. What was the amount of the insurance premiums paid during 1974?
(d) The inventory account balance as of December 31 is $42,850. The employees taking the physical count of inventory report $21,200 of inventory actually on hand at December 31. What adjusting entry should be made?
(e) Mac Company owns three 6% corporate bonds of $1,000 face value. Each bond pays interest annually on June 30. One of the bonds was purchased on September 1, 1974, and the other two were purchased on November 1, 1974. What adjusting entry should be made at the end of 1974?

4-7. RJC Corporation closed its books on December 31, 1974. On January 10, 1975 the auditor for RJC Corporation detected the following errors made in 1974. You have been called in to assist the auditor by determining how each error affected 1974 net income and by what amount and by answering the other questions relating to each error that he has detected.

(a) Depreciation expense of $2,300 was recorded as $3,200. How did this affect assets and by what amount?

(b) Failed to record a $200 expiration of prepaid insurance. Give the amount and effect of this error on assets, liabilities, and owners' equity.

(c) Recorded $350 of interest receivable on a bond held by the corporation as interest expense and interest payable. How did this affect assets and liabilities?

(d) Salaries owed to employees on December 31, 1974 of $400 were not recorded. The employees were paid their proper salaries on January 4, 1975. What sections of the balance sheet were affected by this error?

4-8. The following trial balance and adjusted trial balance were obtained from the records of Wrapup Company on December 31, 1974.

	Trial balance		Adjusted trial balance	
	Dr.	Cr.	Dr.	Cr.
Cash	$ 2,130		$ 2,130	
Accounts receivable	6,850		6,850	
Allowance for uncollectible accounts		$ 640		$ 875
Merchandise inventory	3,000		3,000	
Prepaid insurance			800	
Buildings	16,000		16,000	
Accumulated depreciation — buildings		3,000		4,500
Accounts payable		2,290		2,290
Salaries payable				150
Common stock		18,400		18,400
Retained earnings	210		210	
Sales		23,480		23,480
Sales returns and allowances	230		230	
Cost of goods sold	10,640		10,640	
Freight expense	1,000		1,000	
Salaries expense	6,550		6,700	
Insurance expense	1,200		400	
Depreciation expense			1,500	
Bad Debts expense			235	
	$47,810	$47,810	$49,695	$49,695

REQUIRED

(a) Reconstruct the adjusting entries made on December 31, 1974.
(b) Prepare an income statement for 1974. What observations can you make about the profitability of Wrapup Company prior to 1974? (Omit income tax calculations.)
(c) Prepare a statement of retained earnings as of December 31, 1974.
(d) Prepare a balance sheet as of December 31, 1974.
(e) Close the temporary accounts.

4-9. In the spring of 1975, Kenneth Goodson withdrew $2,000 from his personal savings account and opened a popcorn stand known as the Wildcat Kernel Palace. The stand was located on the campus of a large university and Goodson was given a lease for the months of June, July, and August. The cost of the lease was $300 and supplies for the three months cost $800. Goodson spent $100 on materials to erect the stand, $500 on equipment, and $100 on advertising through the university newspaper. Three students were hired to work at various times throughout the summer at a salary of $100 each. During the three months, Goodson collected $2,425 from the sale of popcorn. The university officials also informed him that he could renew his lease again the following summer. The popcorn stand was no longer usable and had to be torn down. However, the equipment was still in good condition and was estimated to be usable for another four summers. Goodson had supplies in the amount of $75 which could be used the following summer.

REQUIRED. Prepare an income statement for the three-month period and a balance sheet as of the end of August. (Assume Goodson is the sole stockholder in the company and that the stock was originally issued for $2,000.)

4-10. The Regency Repair Company began operations on July 1, 1974. The business was organized by M. I. Sharp and involved the repair of typewriters. Mr. Sharp had a thorough knowledge of typewriters but knew little about accounting records and financial reports. Therefore, all of the financial data for the company were handled by Mrs. Sharp, who had no previous accounting experience. Her major financial records consisted of a checkbook in which she kept the details of all cash receipts and disbursements, a listing of receivables and payables, and a listing of all other assets owned by the business. On December 31, the Company's year end, Mrs. Sharp prepared the financial statements presented as follows:

Regency Repair Company
Assets and Liabilities
December 31, 1974

ASSETS		LIABILITIES	
ash	$ 500	Note payable on truck	$1,500
eceivables from customers	1,250	Payables for spare parts	600
ompany truck	4,000		$2,100
epair equipment	1,000		
	$6,750		

Regency Repair Company
Cash Receipts and Payments
December 31, 1974

Receipts:
Initial investment by Mr. Sharp	$2,500	
From customers	9,420	
Total receipts		$11,920

Payments:
Salaries	$5,920	
Spare parts	800	
Truck	2,500	
Repair equipment	1,000	
Rent for 1 year	1,200	
Total payments		11,420
Gain for the period		$ 500

Mr. Sharp wishes to expand the business, but in order to do so he must secure a loan from the bank. He needs to know whether or not the financial statements as presented here correctly reflect the status of the business for its first six months.

Mr. Sharp estimates that the truck will last five years and the repair equipment will last four years. Both assets will be worthless at the end of this time. There are $400 of spare parts on hand and $130 of unpaid salaries as of December 31. Mr. Sharp owns all the stock of the company.

REQUIRED

(a) Is the company operating at a profit or a loss?

(b) What is the true composition of the company's balance sheet?

(c) Can you recommend a more accurate accounting method for Mrs. Sharp to follow?

4-11. Adjusting entries are made at year end to bring the account balances up-to-date. One characteristic of adjusting entries is that they include one Balance Sheet account and one Income Statement account. Describe a situation and prepare the necessary adjusting entry for the following types of adjustments.

(a) Debit an expense and credit an asset.

(b) Debit an asset and credit an expense.

(c) Debit an asset and credit a revenue.

(d) Debit an expense and credit a liability.

(e) Debit a liability and credit a revenue.

(f) Debit a revenue and credit a liability.

4-12. An inexperienced bookkeeper for the Kossuth Corporation made various entries to the accounts as follows:

(a) A $500 fire insurance bill for the month of December was received on December 30, with payment to be made in January. No entry was recorded.

(b) An addition to the factory building in the amount of $9,000, properly chargeable to the Buildings account, was debited to Repairs and Maintenance Expense. Depreciation expense on buildings is charged at 5% per year.

(c) In recording accrued interest income at year end the bookkeeper made the following entry.

Interest expense	250	
Interest receivable		250

(d) A new machine costing $5,000, purchased on January 2 of the current year, was charged to the Buildings account. Depreciation expense for buildings is 5% per year while depreciation expense on machinery is charged at 10% per year.

REQUIRED. Assume each event is independent of any other. The company year end is December 31. Prepare the journal entry necessary to correct each situation assuming the books have not been closed.

5
CASH

EDUCATIONAL OBJECTIVES

The material in this chapter is designed to achieve several educational objectives. These include:

1. An understanding of the significance of cash management to the success of a business firm.
2. An awareness of the necessity of a good internal control system for all assets to help assure the validity of external financial statements.
3. A knowledge of why a bank reconciliation is important and how a reconciliation should be prepared.
4. An understanding of how a petty cash fund operates and the relationship of a petty cash fund to an internal control system.
5. An appreciation for the fact that internal control systems have common characteristics but must be adapted to the needs of individual firms.

This chapter begins the in-depth discussion and analysis of the major classifications on the balance sheet. As each of the balance sheet items is presented, the pertinent information relative to the income statement will also be discussed. As previously indicated, assets are shown on the balance sheet in their order of nearness to cash; thus the first balance sheet account to be discussed is cash.

In this chapter, some of the discussion of cash will be in the context of what happens within the firm and the activities of the management of the firm. The reason for this is that the reliability of the information on the external financial statements is a function of the internal activities and decisions. The external financial statements are the end product of internal decisions. Therefore, it is important that the user of financial statements have some knowledge of the internal functions.

Definition

The term *cash*, as used in accounting, includes currency and coins, checks from customers, and bank accounts. The importance of cash cannot be overstated because without cash a firm cannot exist. Cash is needed to pay debts, salaries of employees, taxes, etc. The flow of cash permits a firm to invest in assets, produce inventory, and pay dividends to stockholders.

A firm must have sufficient cash to begin operations, but equally important, it must have additional cash several months later. The reason for this is that the items purchased initially on credit must be paid. Therefore, the flow of cash into the firm in the initial months of operation is important to the ultimate success or failure of the firm.

INTERNAL CONTROL

The users of financial statements should recognize that the reliability of the financial data is influenced by the adequacy of the internal control system used by the firm. Evidence of this is the fact that the certified public accountant normally begins the evaluation of the firm's financial data with a review of the firm's internal control system.

According to the American Institute of Certified Public Accountants in the Statement on Auditing Standards Number 1, internal control is:

> The plan or organization and all the co-ordinate methods and measures adopted within a business to safeguard its assets, check the accuracy and reliability of its accounting data, promote operational efficiency, and encourage adherence to prescribed managerial policies.

This definition implies a broad plan of internal control or system. The actual system is left to the discretion of the individual firm. A system applicable for General Motors would not be used, or needed, for a corner grocery store. Likewise, a system to be used in a haberdashery would not fit the needs of U.S. Steel. Each situation must be viewed as unique and a system designed for that particular firm.

The system of internal controls applies to all assets — but the asset with the most apparent need for control is cash. The lack of internal controls with respect to cash probably means that there are no internal controls with respect to anything. If management does not provide controls over cash, then it is almost certain that other assets are not safeguarded. Money can simply be folded up and placed in a pocket if precautions are not enforced. With other assets, e.g., inventory, the size and weight greatly reduce the likelihood of theft. Consequently, the importance of adequate internal controls for cash cannot be overstated.

As an illustration of the effect of inadequate internal controls, consider the following situation which is based on actual events. George Dawkins worked for the Harbor Towing Company in a large eastern seaport city. Harbor Towing had as its principal job the nursing of large freighters and ocean liners into the docks of the seaport town. There was little need for office employees because the majority of the employees worked on the tug boats. There were two principal owners who also were the salesmen. Therefore, George was the only office employee. Since he had worked for the firm for eight years, there was very little fear on the owners' part of George's dishonesty.

George, over time, developed a habit of gambling on sporting events. In the first six months of intense gambling, he won $10,000, which was almost as much as his salary. As time passed, George began to lose. Over the next two and one-half years, George lost and paid $100,000 and owed the gamblers $200,000. To pay off this debt, George was embezzling $6,000 a week from the firm. The embezzling was accomplished through the creation of dummy bills, i.e., payments to firms that do not exist. George would go to a bank, open an account in a corporate name, of which he was the president. When payments were made to this firm, for the dummy bills, he would deposit the checks in the corporate name and then write checks to cover his gambling losses.

The lack of controls is reflected in the fact that:

1. The assets of the company were not being adequately protected which resulted in the loss, by the Harbor Towing Company, of $100,000. In some cases protection may be provided by insurance; however, there was no insurance in this case. In addition, there was no possibility of recovery from George since he did not have any money.

2. The accounting data of the firm are not accurate and reliable since the procedures permitted $100,000 of dummy bills to be processed.

3. The situation indicates that no managerial policies existed relative to the payment of invoices.

The Harbor Towing case illustrates that a system does not have to cost millions of dollars to implement. An effective system for this company could have been as simple as the owners signing the checks. Presumably, the owners would have reviewed the invoices before signing the checks. Consequently, George would have been reluctant and cautious about creating dummy bills. This initial reluctance is an objective of internal controls. A system of internal controls will not necessarily uncover fraud. The system is designed so that people think twice before trying such a scheme. Hopefully this will reduce the chance of fraud and encourage employees to follow prescribed procedures.

The Harbor Towing case is not a unique situation. There are many firms with similar problems. As previously mentioned, each firm has its own unique situation and a system must be designed for that firm; however, there are some basic ideas that are applicable to most situations.

CASH RECEIPTS

The initial objective in the design of a system for the control of cash receipts is to make sure that the physical control of cash is combined, at the earliest possible moment after receipt, with an accurate recording of the amount of the receipt. A journal entry should be made, preferably in a cash receipts journal, as soon as the money is received. As discussed in Chapter 3, a special journal can be designed to minimize the mechanical process of making entries if there are numerous transactions of the same type, i.e., receipt of cash.

Cash Sales

If cash is received at the time of sale, a cash register should be employed. This is typical of department stores where the sales person will run up a cash register tape and collect the money before the customer is permitted to take the merchandise. As a device to encourage the sales person to ring the proper amount on the cash register, a sign stating that "no refunds will be given without a cash register receipt" is usually placed near the cash register. The reason for this policy is that if the amount is not recorded on the cash register, the employee could pilfer the money paid by the customer. The customer is satisfied that the purchase price was paid and is content to walk out of the store. If the money is never placed in the cash register it is, in effect, being given to the employee. Consequently, the policy of cash register receipts before any refunds makes the customer an important control factor in the total cash system.

A check, which is given for the purchase price, presents fewer problems in terms of recording than a cash sale. The check is made payable to the store; therefore, only a poorly written check could be altered by the employee and thus lend itself to fraud.

Collection on Credit

Checks are generally used by customers in paying for merchandise purchased on account. These checks come in the mail and must be recorded as receipts, credited to the proper account, and deposited in a bank. Therefore, a system of internal controls must be employed so that this process is facilitated. The following suggested system would be employed in a large firm but could be modified for a smaller business:

1. The individual responsible for opening mail separates the checks from other types of mail. This person records the number of checks received on a sheet of paper and this number serves as a control. In many systems, this individual will also prepare an adding machine tape of the check amounts, which is sent to the accounting department. The actual envelopes and checks are sent to a second individual.

2. The second individual is responsible for recording the checks on a deposit ticket. The deposit ticket should be prepared in triplicate. This individual keeps a copy and forwards the original and the first copy to a third person.

3. The third person posts the payments to the accounts receivable subsidiary ledger. The checks and two deposit tickets are then sent to a fourth individual.

4. The fourth person reviews the deposit and takes the checks and two copies of deposit tickets to the bank.

5. The final phase of the control process involves a comparison by the accounting supervisor of the duplicate deposit ticket received back from the bank and the initial adding machine tape prepared by the mail clerk. The total and number of items should be the same for both documents.

The preceding system (see Figure 5-1) has five people involved in the receipt of cash. This system would be expensive for many firms because of the labor cost involved. As the size of the firm decreases, fewer people would be available for separation of duties. Nonetheless, even with fewer people, there must be some controls. At a minimum, two people should be involved.

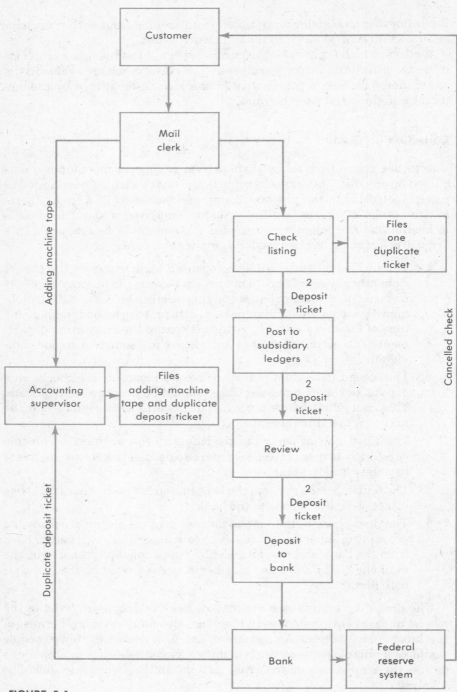

FIGURE 5-1

Cash receipts system

Basic Principles

There are some basic rules that should be followed in any system for controlling cash receipts:

1. There should be as much distribution of responsibilities as possible.
2. All receipts should be deposited in the bank on the day received. There should be no exceptions to this rule.
3. In larger firms, the internal audit staff should give special attention to cash receipts through surprise audits.
4. When procedures have been established, management should insist that they be followed.

CASH DISBURSEMENTS

Whereas the control of cash receipts ties in with the cash receipts journal, the control of cash disbursements is related to the cash payments journal. The activities associated with cash disbursements are under the control of management, in contrast to cash receipts where outsiders decide the method and time of their payment.

The major objectives in the control of cash disbursements are to preclude payments that are not justified and to support with evidence those obligations which are to be paid. The Harbor Towing Company case illustrates the failure to achieve these objectives. Payment was made to firms that had not supplied any merchandise and the evidence, used to support the payment, was invalid.

Basic Principles

The division of responsibilities is as important for cash disbursements as for cash receipts. Several employees should be involved in the disbursement process. The checks, used for payment, should be serially numbered and tightly controlled. There should be supportive documents before a check is prepared, i.e., purchase order, receiving report, inspection report, and invoice from supplier. These documents are reviewed and serve as the basis for the preparation of a voucher.

In a voucher system, a voucher is a special form on which is recorded the data about the transaction, for example, the payee, invoice number, accounts to be debited and credited, amount purchased, and any adjustments. Once the information is put on the voucher, the individual responsible for authorizing payment reviews the total package of documents and, if everything is in order, signs the voucher indicating authorization for payment.

Cash Disbursement System

The system presented is elaborate, but as was mentioned with the cash receipts system, it can be modified to fit particular firms. A major control point is that no one connected with the cash receipts system should be a part of the cash disbursements system. If this is not possible, then there should be close supervision of those responsible for cash receipts and disbursements.

The more elaborate system would include:

1. An individual responsible for receiving the data connected with each invoice, i.e., receiving reports, inspection reports, delivery bills, etc. As the due date approaches, the various items connected with each invoice are accumulated and a voucher prepared. This individual records the voucher number and amount on a separate control sheet.

2. The completed voucher, i.e., purchase order, receiving report, inspection report, invoice from supplier, with the voucher form on top, are reviewed by the individual responsible for authorizing payment. As previously mentioned, if everything is in order, this individual signs the voucher indicating authorization for payment.

3. The voucher and supplemental materials are then given to a third person who prepares the check. The checks are also prenumbered and can be prepared in duplicate to provide additional controls over disbursements.

4. The voucher and the check are then reviewed by an individual who has authority to sign the check.

5. After the first signature, the entire package is given to another person with authority to sign the check. The check is signed, after review, and given to another individual who actually mails the check.

6. After the checks are mailed, the entire package is given to another person who posts the amounts to the accounts payable subsidiary ledger.

The foregoing system (see Figure 5-2) is obviously ideal for large firms with sufficient volume to warrant the personnel; however, the system described contains the basic ideas which should be used by all firms irrespective of size. In addition, the increased use of computers and the ability to control through the use of a centralized system have enabled firms to reduce processing costs.

To review, some of the basic concepts of any cash disbursement system are:

1. Division of responsibilities for the several activities involved in cash disbursements.

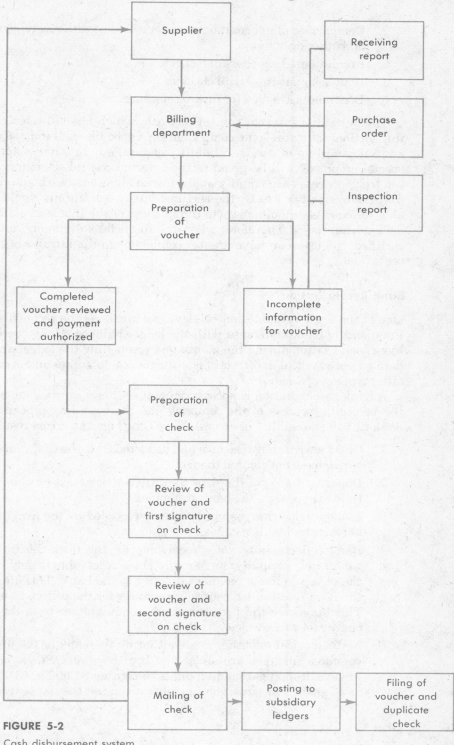

FIGURE 5-2

Cash disbursement system

2. Comparison of information on seller's invoice with receiving reports and other documents.

3. Prenumbered vouchers and checks.

4. Double signatures on all checks.

5. Documents are filed for future reference.

The system of internal control for cash, which has been described in this section, indicates some components of a good internal control system, i.e., safeguarding of assets, reliability of data, etc. In other chapters, no special emphasis will be given to the internal control system; however, the basic concepts are equally applicable to all assets. Without adequate controls over other assets, the certified public accountant would not be able to express an opinion as to the fairness of the financial statements for external users. Therefore, when the financial statements have been certified, the user can have greater confidence in the fairness of the presentation.

Bank Reconciliation

One of the most valuable control devices for cash is the reconciliation of the monthly bank statement with the ledger balance. Many checks may have been written during the month and reconciling the ledger and bank balances is essential in correcting mathematical mistakes and preventing other types of errors.

A bank reconciliation is necessary because there is a time lag between the recording process of the company and the recording process by the bank and the possibility of errors. These time lags and errors result from:

1. Checks written by the firm and deducted from the cash account but which have not cleared the bank.

2. Deposits made by the firm that are not recorded by the bank in time to appear on the bank statement.

3. Bank service charges not known or recorded by the firm until the bank statement is received.

4. Bank collections of notes receivable for the firm. Since the firm would not, normally, have many notes receivable it may instruct the payee to make payment directly to the bank. This eliminates the need to establish control procedures for the collection of notes. This may save the firm time and money greater than the bank's charge for this service.

5. Mathematical mistakes made either by the bank or the firm. Two common mistakes are slides and transpositions. *Slides* normally involve dropping or adding digits (recording $1,000 as $10 or $200 as $2,000). *Transpositions* involve reversing the digits (recording

56 as 65 or 92 as 29). These are common errors and care must be exercised in the initial recording process.

As an illustration of the preceding items, and the preparation of a bank reconciliation, consider the following data:

Balance per bank statement, July 31 $14,500
Balances per books, July 31 $13,410

Other information:

1. Outstanding checks, July 31 (verified as correct). $ 1,180
2. Deposits in transit.   1,000
3. Customer check returned with bank statement because of insufficient funds (these checks are called *NSF*, meaning not sufficient funds).   60
4. Service charge by bank (recorded by bank but not by firm).   5
5. Error by firm (check for $72 recorded by firm as a $27 payment on account).
6. Note collected by bank for firm (includes $20 interest).   1,020

XYZ COMPANY
Bank Reconciliation
as of July 31

Balance per bank	$14,500	Balance per books		$13,410
Add:		Add:		
Deposit in transit	1,000	Note collected by		
	$15,500	bank (includes		
		interest)		1,020
				$14,430
Deduct:		Deduct:		
Outstanding checks	1,180	NSF check	$60	
		Service charge	5	
		Error in		
		recording	45	110
Adjusted balance	$14,320	Adjusted balance		$14,320

Note that the adjusted balances are equal. Given all the information, these balances must be equal since any difference must be located. Banks normally allow ten days before the balances, as presented by the bank, are assumed to be correct. Consequently, the reconciliation should be made as soon as the bank statement is received.

There are two additional points to be discussed concerning the bank reconciliation:

1. The NSF check is a deduction from the balance per books because the check was originally deposited by the firm, thereby increasing the cash account balance. The bank increases the firm's account pending collection. If collection is made no further action is necessary. If collection is not made, then the bank reduces the firm's account and returns the check to the firm. Collection on the check is the responsibility of the firm accepting the check, not the bank.

2. Journal entries must be made for all items that affect the balance per books. These items require entries because they have not been previously recorded. Adjustments to the balance per bank should not be recorded by the firm. These items affect the bank records, not the records of the firm. The adjusting entries for the example would be:

Cash	1,020	
Notes Receivable		1,000
Interest Income		20
Accounts Receivable	60	
(NSF check)		
Bank Service Expense	5	
Accounts Payable		
(error in recording)	45	
Cash		110

Note that the NSF check becomes a receivable since the check was returned and payment was actually not received. Accounts Payable is reduced by the amount ($72 − 27) of the transposition.

PETTY CASH FUND

Obviously, installing a complex cash disbursement system is expensive. The cost of processing a voucher includes the cost of material, the overhead (heat, light, power, etc.), and most importantly, the labor cost of the time spent by each individual. Consequently, it is as expensive to process a voucher for 50 cents as to process a voucher involving thousands of dollars. The firm would desire substantial control over the large purchases, but in view of the cost, it might be willing to accept less control over smaller purchases. The cost of processing the voucher for a small purchase might be greater than the purchase price. As a result, many firms operate petty cash funds. A petty cash fund is created for the payment of the many small purchases of the firm.

Management will select an employee to be the custodian of the fund. The size of the fund is a function of the amount to be used, generally, over the course of one month. Amounts that do not cover at least a month's activity necessitate frequent replenishments and thereby minimize the value of the fund.

Establishing the Fund

A check from the regular cash account is prepared, payable to the custodian. The check is cashed and the money is placed under the control of the custodian. A list of the items acceptable for payment out of the petty cash fund should be prepared. For example, stamps, small office supplies, tolls, and parking fees. Such items as loans to employees should not be permitted.

An entry must be made to record the transfer of cash from the regular cash account to the petty cash fund. The entry for the creation of a $250 fund would be:

$$\text{Petty Cash} \quad 250$$
$$\quad \text{Cash} \quad\quad\quad 250$$

The procedure for distributing cash for the various purchases should require the recipient to sign a slip indicating the amount received, nature of expenditure, and name of recipient.

Replenishing the Fund

When the amount of cash has been reduced to a specified level, the custodian will initiate a voucher. The slips will be grouped by category according to the prescribed chart of accounts and delivered to the individual responsible for preparing vouchers. The list of expenses will be reviewed, and at the conclusion of the voucher process, a check issued for the replenishment of the fund. Utilizing this process, numerous minor items are combined before a voucher is processed. This will result in significant savings in time and labor cost and will outweigh any minor losses incurred by having the petty cash fund.

An entry is made for the various expenses that constitute the voucher. As an example, if a check for $175 is prepared to replenish the fund, the distribution of the items on the voucher might be:

Postage Expenses	18
Office Supplies	70
Entertainment Expenses	35
Travel Advances	40
Miscellaneous Expenses	12
Cash	175

Any small difference between the total receipts and the amount necessary to replenish the fund is handled through a miscellaneous account entitled Cash Over and Short. Therefore, in the petty cash fund there will always be $250. This total is either in the form of cash or slips of paper representing amounts paid for minor items. On the balance sheet, petty cash in the amount of $250 may appear separately or be included in the cash balance.

BALANCE SHEET PRESENTATION

Cash on Hand, Cash in Bank, and Petty Cash are included in the Current Asset section of the balance sheet. They are usually grouped together to prevent this section from being cluttered by numerous small balances.

Certain cash-type items do not appear in the Current Asset section. Any sum of money set aside for special purposes, i.e., a fund established to retire a bond issue, is generally not included in the Current Asset section. These items are not going to be used to pay current obligations; therefore, they should not be classified as current assets.

A typical balance sheet for cash and a fund to retire bonds would be:

Current Assets:

Cash (including petty cash of $250) $51,500

Investments:

Fund to retire bonds   25,000

MANAGERIAL PROBLEMS OF CASH

The management of cash is the major consideration in achieving the desired liquidity position of the firm. The paradox of liquidity–profitability is the fundamental problem in cash management. This paradox arises from the fact that the more cash on hand, the easier it is to pay obligations which come due. However, more cash available to pay bills means fewer dollars invested in assets, thereby reducing profitability. Therefore, the optimum level of cash for the firm must be determined.

In developing the optimum level all factors connected with cash must be considered. Among these items are:

1. *Predictable discrepancies between the inflows and the outflows.* Not all expenses are paid every month: short-term borrowings are repaid at varying times; fixed assets may be paid for at the time of purchase or payment may be spread out over time; dividend payments are made quarterly, semiannually, or annually. Transactions of this type necessitate planning and providing for cash to meet payment obligations.

2. *Unpredictable discrepancies between inflows and outflows.* There are emergencies which occur that are not planned: failure of customers to settle their accounts; strikes; and floods. These are events that, although not foreseen, must be provided for by the firm.

3. *Availability of methods to speed collections and/or delay payments on bills.* Management should be concerned with getting cash in as soon as possible. A credit society demands time payments and as a result, there is a delay between the sale and the collection of cash. In some cases, management can influence the flow of cash by offering inducements for prompt or early payments. The nature of these inducements will vary with the firm.

    On the other hand, management should attempt to postpone payments to others for as long as possible, within the bounds of reasonableness. This is especially true if there is no interest attached to the obligation. If there is interest, then the benefit of delaying payment must be weighed against the interest cost incurred by nonpayment.

CASH BUDGETING

The planning of cash inflows and outflows involves the development of a cash budget. The idea of a cash budget is not unique to business firms. Each individual is faced with the same problem — trying to stretch limited dollars to cover seemingly unlimited desires. The widespread use of credit cards has made the purchase of goods so easy that the self-discipline inherent in a budget has been made more difficult. Business firms share this same dilemma — buying without having sufficient resources to pay for the purchases. Therefore, to facilitate the achievement of the desired objectives of the firm, a sound and logical budget must be prepared.

A cash budget is an attempt to predict the flow of cash in and out of the firm over a particular time span. This is basically the same as an individual attempting to predict the various entries in a personal checkbook. The objective is to determine whether there will be sufficient cash available to meet expected obligations and, if not, when it will be available and how much money will be needed. If, on the other hand, there will be more money than needed at any one time, the amount of the excess should be determined and consideration given to possible uses for the excess.

The excess cash may be either temporary or permanent, i.e., long-term. In the case of a temporary excess, the cash should be invested so that it will be available when needed, e.g., in short-term government notes. A permanent excess could be invested in assets which will contribute long-run income to the firm, e.g., machinery and equipment.

Temporary Excess

Assume that a firm has an excess of cash inflows over cash outflows of $50,000 in February, but it will have a like deficit in June. Management could invest these funds in short-term notes in order to have them available to meet the expected deficit. Assume that the interest rate on these short-term investments is 5% per year or $1\frac{1}{4}$% for the 90 day period from March 1 to June 1. Thus, the firm could expect to receive $625 in interest ($50,000 $\times$ $\frac{5}{100}$ $\times$ $\frac{1}{4}$). This amount of interest may not appear to be significant, but it is some return on the idle funds. Without a cash budget, management might not have anticipated the temporary excess; consequently, they would not have made plans to invest these funds.

Permanent Excess

A permanent excess would be invested in those areas, e.g., higher inventory levels, new machinery, etc., which should generate a significantly higher return than that earned by short-term investments. Rates of return that are acceptable vary with each firm because managers have different concepts regarding profit goals. The analysis of these potential long-term investments is much more detailed and complex than for short-term government obligations. This process is called *capital budgeting* and is discussed more fully in Chapter 15.

Preparation of a Cash Budget

The first step in the preparation of a cash budget is the selection of the time period to be covered. This period could be six months, one year, five years, etc. The longer the period of time, the more difficult the preparation of the budget and the more likely it is to be wrong. This results from the fact that many estimates must be made and carrying these estimates out for too long a period of time will produce inferior results.

As an example of forecasting, the receivables collected in a month are derived from previous month's sales. Percentage estimates must be used since all receivables are not collected in the month after a sale. Sales in May of $90,000 may be collected 70% in June, 20% in July, and 10% in August. Therefore, the cash inflow in June will be only $63,000 with the balance collected in succeeding months, $18,000 in July and $9,000 in August. These estimates, and others required for the preparation of a cash budget, are extremely difficult to make, and the use of mathematics does not increase the validity of poor initial estimates.

XYZ COMPANY
Cash Budget
for the period 7/1/74–12/31/74

Receipts	July	August	Sept.	Oct.	Nov.	Dec.
Collections on account	$ 9,600	$11,000	$15,400	$18,200	$22,400	$23,000
Payments						
Wages paid	2,200	2,200	2,400	2,400	2,700	2,800
Payments on account	7,000	8,200	7,500	7,000	8,000	8,400
Direct factory costs	1,700	1,800	1,800	1,800	2,000	2,000
Administrative expenses	800	800	800	900	900	900
Selling expenses	1,000	1,100	1,100	1,200	1,400	1,400
Purchases of equipment			4,000			2,000
Dividends paid				2,000		
	$12,700	$14,100	$17,600	$15,300	$15,000	$17,500
Excess of inflows over outflows	(3,100)	(3,100)	(2,200)	2,900	7,400	5,500
Cash balance at beginning of month	1,500	400	300	100	1,000	3,400
Expected cash balance at end of month	($ 1,600)	($ 2,700)	($ 1,900)	$ 3,000	$ 8,400	$ 8,900
Plus: Amount of loan necessary to cover deficits[a]	2,000	3,000	2,000			
Less: Loan repayment				2,000	5,000	
Cash balance at end of month	$ 400	$ 300	$ 100	$ 1,000	$ 3,400	$ 8,900

[a] Cash will be borrowed through a credit line during the month to cover expected deficit. In this example, the amount borrowed must be in multiples of $1,000. This example has been simplified to illustrate the point. A more detailed look cash budgets will be covered in either a cost accounting course or a course in finance.

FIGURE 5-3

A cash budget reflects cash inflows and outflows; therefore, any item that does not result in a cash inflow or outflow is not included in the cash budget. The information for the budget is developed from many sources. The inflow information comes from the sales department, which will have projected the amount of sales. The accounting department, in preparing the cash budget, will then take into consideration the expected delays in the collection of sales. The cash outflow information must be obtained from all departments, e.g., sales, payroll, and general accounting records.

A typical cash budget is presented for illustrative purposes (see Figure 5-3).

SUMMARY

The control of cash is a major managerial responsibility. The accountant assists management through the preparation of cash budgets and the establishment of sound internal control systems for receipts and disbursements. The systems may range from simple to highly complex depending upon the size and nature of the firm.

QUESTIONS

5-1. The accountants' major role in accounting for cash involves developing a sound internal control system. What is the most important feature of a sound internal control system? Is a system of internal control over cash possible in a small company employing only one bookkeeper?

5-2. Because the accountant is aware of all the financial transactions which occur in a business, it is possible for him to develop an accounting system in which embezzlement and fraud are impossible. Comment on this statement.

5-3. In a department store where a cash register is used, how is the customer an important control factor in the total cash system?

5-4. List the basic rules to be followed in a system designed to control cash receipts.

5-5. What is the major objective in the control of cash disbursements? Is this objective equally obtainable in both the small and large firm? Explain.

5-6. Is the division of responsibilities as important in the area of cash disbursements as it is in the area of cash receipts? Explain.

5-7. The accounting clerk for ABC Company was instructed by the chief accountant to obtain a suppliers' invoice, receiving report, inspection report, and bill from the supplier before making any cash disbursements.

The clerk believes that the supplier's invoice and bill are sufficient evidence to warrant payment. What reasons can you suggest for requiring the other two documents?

5-8. List the basic rules to be followed in a system designed to control cash disbursements.

5-9. At the end of each month the bank sends a statement which lists all the transactions they have processed for the firm. The balance shown on the statement rarely equals the balance of a company's Cash account. What are the reasons for the discrepancy?

5-10. Explain how each of the following items would be reflected on a bank reconciliation:
(a) Outstanding checks of $485.
(b) NSF check from customer returned with bank statement.
(c) The bank recorded a deposit of $980 as $890.
(d) Bank sent notice to company that its request for a $5,000 loan had been accepted and the money added to the firm's account.
(e) Service charge for the month amounted to $25.

5-11. Which items on a bank reconciliation require journal entries? Do these entries always represent correction of errors? Explain.

5-12. What is the purpose of establishing a petty cash fund? Why is the account Petty Cash not included in the entry for the reimbursement of the fund?

5-13. What is the liquidity–profitability paradox concerning cash?

5-14. What are some important factors involved in determining the optimal level of cash for a firm.

5-15. The Citation Company has earned a profit in each of its first five years of operations. During this period the company's bank account has steadily increased. Do you believe the company is gaining optimal benefits from its cash reserves? If not, what suggestions would you make?

5-16. What elements comprise a cash budget? Why is a cash budget prepared for a short period of time superior to one prepared for a long period of time?

PROBLEMS

Procedural Problems

5-1. January 1, the Huskie Company had an accounts receivable balance of $135,000, which originated from credit sales made during the previous three months as follows:

Month	Credit sales
December	$100,000
November	90,000
October	85,000

The company's past experience shows that approximately 70% of the credit sales of any month are collected in the following month. An additional 20% are collected in the second month following the month of sale, and the remaining 10% are received from customers in the third month.

REQUIRED. If credit sales are estimated to be $105,000 in January, calculate the amount of cash Huskie Company can expect to collect from credit sales in the months of January and February.

5-2. At the end of 1973, the accountant for the Zander Company projected cash transactions as follows:

	January	February	March
Cash sales and collection of receivables	$32,000	$60,000	$65,000
Merchandise purchases	18,000	40,000	50,000
Salaries expense	8,500	8,500	10,500
Annual rent		3,600	
Insurance premiums	6,000		
Addition to plant		15,000	

The January 1, 1974 cash balance is $1,700. If cash is expected to be less than $1,000, the company borrows funds from the bank. Loans are in multiples of $1,000, and the company desires to maintain at least a $1,000 cash balance. Payments will be made to the bank in multiples of $1,000 when cash is above the minimum requirement.

REQUIRED. Prepare a cash budget for the first three months of 1974.

5-3. Each of the following independent cases refers to the bank account of Gardner Company:
   (a) The bank statement for the Gardner Company at March 31, 1973 shows a balance of $71,200. Outstanding checks are $39,300; deposits in transit are $7,600. Based upon these data, compute the total amount of checks that can be authorized for issuance without overdrawing the bank account.
   (b) As of July 1, the Gardner Company had outstanding checks of $119,000. During the month the company issued $274,000 in additional checks. The bank statement as of July 31, shows that $311,400 of checks cleared the bank. Compute the amount of outstanding checks as of July 31.
   (c) On October 1, Gardner Company had outstanding checks of $17,000. During the month the president approved and issued $174,000 of additional checks. The bank statement of October 31 shows deductions of $169,000; included in this figure is a bank service charge of $45. Compute the checks outstanding as of October 31.
   (d) The bank statement of December 31 was reconciled with the book balance as of that date. In the process of examining the bank reconciliation, the auditor for Gardner Company discovered that a check from a customer, which had been deposited, was returned with the bank statement. In reconciling the bank statement, the company accountant had added this check to the bank balance in arriving at the

adjusted cash balance. How should this item have been handled? Why?

5-4. On March 1, 1973, the Hanks Company established a petty cash fund of $400. On April 30, the composition of the fund was as follows:

Coins and currency	$ 95.20
Postage stamps	25.00
Delivery charge invoices	142.50
Office supplies receipts	84.80
Salesmen expense advances	52.50

REQUIRED. Prepare the entries to:
(a) Establish the fund.
(b) Replenish the fund on April 30.
(c) Increase the fund by $100.

5-5. The bank statement for Big Ben Company indicates a balance of $7,169.33 on March 31, 1974. The Cash account shows a debit balance of $6,993.50 after the cash journals have been posted for March. Comparison of the bank statement, and the accompanying checks and bank memos, with the books reveals the following:
1. Checks outstanding totaled $2,743.04.
2. Among the paid checks returned by the bank was a check for $83.71 drawn by Big Bill Company.
3. The bank had collected a note for the Big Ben Company in the amount of $830 ($800 principal, $30 interest).
4. The bank charged Big Ben Company $3.50 for service charges during March.
5. A deposit on March 31 for $3,065 was made too late to appear on the bank statement.
6. A check for $260 returned with the bank statement was recorded in the books at $235. The check was for the payment of an obligation to Krupp Company for office supplies purchased on account.
7. Among the checks returned was one received from M. I. Broke on account in the amount of $220. His bank had stamped NSF on the check.

REQUIRED.
(a) Prepare a bank reconciliation as of March 31, 1974 for the Big Ben Company.
(b) Prepare all necessary journal entries to correct the books of Big Ben Company.

5-6. The following information pertains to the bank reconciliation of Becherer Company as of May 31, 1974:
1. Balance of the Cash account on Becherer Company's books as of May 31, 1974 was $8,931.40.
2. Balance shown on the May 31, 1974 bank statement was $8,365.90.
3. One of the checks returned with the May bank statement was stamped NSF. The check was originally received from S. Mathews for payment on his account in the amount of $150.00.

4. The bookkeeper for Becherer Company reported outstanding checks totaling $1,361.50; however, your investigation revealed the following:
   - Check no. 263 was listed on the outstanding check list as $643.00, but it was actually written and recorded for $436.00.
   - Check no. 281, in the amount of $93.00, was returned by the bank but misplaced by the company bookkeeper.
5. Also accompanying the bank statement was a cancelled check of Beeher Company for $325.00; the bank had deducted this check from the Becherer Company account.
6. A deposit on May 31 for $1,641.00 was made too late to appear on the bank statement.
7. On May 30, 1974, the bank collected a note for Becherer Company in the amount of $540.00 (principal $525.00, interest $15.00). The bank charged a collection fee of $3.20.
8. Bank charges for the month were as follows:
   - Printing checks — $35.00
   - Service charge — $12.80

REQUIRED. Prepare a bank reconciliation for Becherer Company as of May 31, 1974. Also, prepare the journal entries necessary to correct Becherer Company's books on May 31, 1974.

Conceptual Problems

5-7. The Hartman Company established a $500 petty cash fund on October 1, 1973. On October 31, 1973, the accountant for Hartman Company made the following journal entry:

Entertainment Expense	80	
Delivery Expense	45	
Supplies Expense	110	
Postage Expense	50	
Cash Over and Short		6
Petty Cash		150
Cash		129

REQUIRED. Using this information answer the following questions:
(a) What did the credit to the Petty Cash account accomplish?
(b) What was the amount of cash in the petty cash fund immediately before the preceding entry was made?
(c) What is the reason for including the account Cash Over and Short in the preceding entry? On what financial statement does the account appear?
(d) Assuming Hartman Company prepares interim financial statements at the end of each month, is it important that the petty cash fund be replenished at the end of each month? Explain.
(e) On November 3, 1973, $50 was taken from the petty cash fund and given to a salesman for expenses he had incurred. What journal entry should the petty cash fund cashier make on this date?

5-8. The Clay Company began operations on January 1, 1974. The Company policy was to make all disbursements by check and deposit all receipts daily. As of January 31, 1974, the cash balance per bank statement and balance per books were $1,450 and $1,275, respectively. This difference resulted from outstanding checks which amounted to $410 and a deposit in transit of $225. The bank had also charged Clay Company a $10 service charge for the month of January which was not recorded on the books until early February.

The accountant for Clay Company is now trying to reconcile the Cash account for the month of February and asks for your help in calculating outstanding checks and deposits in transit as of February 28, based upon the following information:

Receipts and payments recorded by the bank during February were $5,780 and $4,950, respectively. Receipts and payments recorded by Clay Company during February were $5,100 and $4,720, respectively. The payments recorded on the books during February included the $10 service charge for January. Included in the bank receipts is a note collected for Clay Company which was not recorded on the books. The bank payments include $25 service charge for the month of February, not recorded on the company books. As of February 28, the balances per bank statement and per Clay Company books were $2,280 and $1,625, repectively.

REQUIRED

(a) Calculate the outstanding checks and deposits in transit as of February 28. (Assume all outstanding checks and the deposit in transit shown at the end of January cleared the bank during February.)
(b) Prepare a complete reconciliation for February.
(c) Give the entries necessary to correct Clay's books as of February 28.

5-9. The Lish Company has been in business for ten years. During the period, the company's profits have risen at a steady pace. The accounting department has four employees.

W. Starr:	Chief accountant
J. Dyer:	Petty cash custodian and secretary
L. Domer:	Payroll clerk
J. Talbott:	Accounts receivable and payable clerk

The chief accountant's function is to supervise the accounting department as well as to serve as financial vice-president. This latter position consumes the majority of time; however, since the other three employees have been with the company since its inception, Starr believed their functions needed only minimal supervision.

The extent of Starr's supervision was (a) the audit of the petty cash fund once a month and (b) issuing the check to replenish the fund personally. The company employed only 40 people; therefore, he only reviewed payroll checks weekly to determine that the proper people were included and paid their authorized wage. His control procedure for accounts receivable was to review all accounts on a monthly basis and to mail past due account notices personally. The control he exercised over accounts payable was a review of the invoices and supporting documents before signing the checks. He then returned these data to Talbott who filed the records, prepared the entries, and mailed the checks.

During the most recent audit, fraud which cost the company an estimated $50,000 was discovered. The fraud was perpetrated by Talbott in the areas of receivables and payables. When checks were received for accounts receivable, he would deposit a portion of the proceeds in his personal account and record a credit to the customer's account. The major source of the fraud in the area of accounts payable was that a major supplier of materials for the Lish Company was owned by a friend of Talbott's. Talbott would change the date on a previously paid invoice and submit it together with the supporting documents for repayment. Talbott and his friend would then split the proceeds. As a result of the fraud, Talbott was fired and a new clerk hired. The chief accountant now asks you to aid him in designing a system, utilizing existing personnel, to prevent a recurrence of fraud.

5-10. The following bank reconciliation was presented to you during your review of the accounts of Laval Corporation at December 31, 1974:

Cash on hand per books		$168,450.00
Less:		
Sinking fund for bond retirement	$21,200.00	
Employee pension fund contribution	15,500.00	36,700.00
Free cash balance		$131,750.00
Add:		
Checks written in 1974, not presented for payment	$ 9,400.00	
Checks written in December, 1974, not sent to creditors until January, 1975	12,650.00	22,050.00
		$153,800.00
Less:		
Payments from customers received on December 31, 1974, not deposited until January 2, 1975	$23,400.00	
Loans from company officers, contracted for in mid-December, received in early January	65,000.00	
Bank charges	2,500.00	90,900.00
Bank balance, December 31, 1974		$ 62,900.00

REQUIRED
  (a) Comment on the reconciliation as presented. If you feel some of the accounts included in the reconciliation are improperly classified, indicate their correct classification.

(b) What is the actual cash balance to be shown on the December 31, 1974 balance sheet of Laval Corporation.

5-11. The Peterson Company, a large midwestern manufacturer, has been in operation for five years. During that period, the company has experienced rather modest growth. The major shareholder, John Peterson, was concerned about the lack of growth so he hired a team of auditors to investigate the situation. The auditors discovered that an employee, Frank Drab, had manipulated the company bank accounts each year to cover up a fraud. The company had three bank accounts with different midwestern banks. During the year, Drab would convert company cash to his own use. To cover up the shortage, he used the following procedures. At each year end he would draw a check on two of the three bank accounts and deposit the amounts in the third account. Because this process was done at year end the banks from which the cash was drawn did not record the transaction until the beginning of the new year. This made the total amount of cash appear greater at year end than the company's actual balance. By the time the checks cleared the accounts, the financial statements were prepared with the larger cash balance.

REQUIRED

(a) Discuss the procedures you would have used to detect this error had you been called in to make the investigation.

(b) What procedures would you recommend to company officials which would prevent a similar occurrence in the future?

5-12. Maronick Corporation has a large balance in its cash account. The account balance is approximately $125,000 in excess of budgeted cash needs for the coming year; however, the company management is reluctant to pay the excess cash out as a dividend due to an expansion program they contemplate undertaking in the following fiscal year. Thus it has been suggested that the company invest the excess funds as follows:

- Place the funds in a savings account at 4½% interest.
- Purchase medium grade corporate bonds which yield 5¼% interest per year and mature in 20 years. The bonds are currently selling at maturity value, i.e., a $1,000 bond is selling for $1,000. Brokers' fees for buying and selling the bonds may be ignored.
- Purchase common stock of Sheffield Company. The stock is presently selling at $50 per share, and its most recent yearly dividend amounted to $3 per share. Over the past 12-month period the stock price has ranged from $41 to $62. Brokers' fees for purchase and sale may be ignored.

REQUIRED

Analyze the three alternatives for investing the excess funds, indicating both the positive and negative aspects of each. List any additional information you would require before making a final decision.

6

RECEIVABLES

EDUCATIONAL OBJECTIVES

The material in this chapter is designed to achieve several educational objectives. These include:

1. A knowledge of the criteria for determining the existence and type of receivable
2. An understanding of the problems involved in the timing of revenue recognition and its effect on the income statement and the balance sheet.
3. A knowledge of the alternative methods of determining the provision for bad debts and how the choice of a method affects net income and the balance sheet valuation of receivables.
4. An insight into the components of a typical system for accounts receivable.

The term *receivable* refers to a claim that one entity has against another. There are different types of claims and, therefore, different types of receivables. The emphasis in this chapter is on accounts receivable and notes receivable, but the concepts discussed are applicable to any type receivable. Other receivables would include dividends and interest.

ELEMENTS OF A RECEIVABLE

The essential elements required before a claim can be considered a receivable are:

1. The amount of money which the firm can expect to receive must be known with reasonable certainty.
2. The due date can be reasonably estimated.

The first element refers to the valuation of receivables. A firm wants to collect every receivable, but it cannot reasonably expect that all claims will result in collections. If the firm learns that the receivable will not be collected, then it is not an asset and must be removed from the accounts.

A second factor to be considered in valuing receivables is the length of time until collection. A sum of money to be received in the future does not have the same value as a sum of money received today. The longer the time interval between the sale and collection, the greater the difference between the discounted present value and the maturity value. The time interval for an account receivable is considered to be short, and because of this, the assumption is that the present value and maturity value are equal. However, with a note receivable, the time interval is longer and an interest rate, to compensate for the time value of money, is generally attached to the note.

A noncash sale of goods and/or services that the firm is in business to sell which meets these two conditions is considered to be an *accounts receivable.* If the receivable arises from some source other than the sale of goods and services it is not an accounts receivable. In that case, the receivable will typically be either a note, interest, or dividend receivable and each should be separately recorded.

ACCOUNTS RECEIVABLE

Accounts receivable from customers generally represent a substantial portion of the firm's current assets. Therefore, the accounting procedures for receivables and the maintenance of adequate control over the extension of credit and methods of collection assume major significance. A poor screening of credit applicants and haphazard collection methods can result in reduced cash inflow, which could require the firm to seek additional capital to finance sales. On the other hand, sound management of receivables can result in the improvement of operations. With the increased usage of credit in our society, the management of receivables increases in importance.

Accounting for Accounts Receivable

There are two primary objectives in accounting for accounts receivable:

1. Determination of the valuation of accounts receivable to be shown on the balance sheet.
2. Determination of the amount of bad debt expense to be shown on the income statement.

Thus there are valuation and matching problems. The accounting procedures may emphasize either valuation or matching. For example, the accountant may approach the situation from the point of view of arriving at a valuation for the balance sheet with the income statement amount of secondary importance. On the other hand, the approach may be to determine the portion to be shown on the income statement with the balance sheet valuation the residual amount.

Revenue Recognition. Another aspect of accounting for accounts receivable is that the gross amount of receivables is in part dependent on the timing of revenue recognition. On the accrual basis of accounting, revenue is generally recognized at the point of sale rather than when cash is collected. Therefore, a receivable is created whenever the recognition of revenue precedes the collection of cash. Thus the balance sheet valuation of receivables is also inherently related to the revenue recognized in the income statement.

Revenue is most commonly recognized at the point of sale; however, there are some exceptions which present certain accounting problems. For example, when should revenue be recognized on a construction project that will take three years? There is precedent in accounting to recognize the revenue on such projects on the basis of the percentage of completion in each year or at the completion of the contract.

Another exception would be recognizing revenue only at the time of the collection of cash. This method can be used in those situations when it is impossible to estimate the amount of the receivable that will be collected. These exceptions to the recognition of revenue at the time of sale are rather specialized; therefore, throughout this text revenue will be assumed to be recognized at the point of sale.

Sales Entry. The entry for the actual sale is a debit to Accounts Receivable and a credit to Sales. As demonstrated in Chapter 3, this entry can be made in a special journal called the sales journal or in the general journal. Assuming a sale of $500 on account, if the general journal is used, the entry would be:

$$\text{Accounts Receivable} \quad 500$$
$$\text{Sales} \qquad\qquad\qquad 500$$

A debit to Accounts Receivable increases that account. The account will subsequently be decreased when payment is received, and the entry for the collection of the receivable is:

$$\text{Cash} \qquad\qquad\qquad\qquad 500$$
$$\text{Accounts Receivable} \qquad\quad 500$$

Sales Returns and Allowances

Some of the merchandise shipped to customers may be either the wrong item (weight, color, or size) or damaged in transit. When this happens, the customer will not pay the full price. The adjustment that is given to the customer must be recorded in the journal. The following entry for an adjustment of $50 could be made:

$$\text{Sales} \qquad\qquad\qquad\qquad 50$$
$$\text{Accounts Receivable} \qquad\quad 50$$

This is not the entry that is normally made. If the firm recorded the above entry there would not be a separate record in the accounting system of the amount of customer adjustments for the period. The firm would only have the net sales that were made (net sales meaning gross sales less sales returns and allowances). Information concerning customer adjustments is important to the firm as a control device over the marketing phase of operations. If the sales returns and allowances are high, customer dissatisfaction can be expected and a potential loss of sales may exist. Because information on sales returns and allowances is important, a special account is used in order that this information will be available to management. Therefore, the entry for the adjustment of $50 of the purchase price would be:

$$\text{Sales Returns and Allowances} \quad 50$$
$$\text{Accounts Receivable} \qquad\qquad 50$$

The debit to the Sales Returns and Allowances account reduces gross sales to the net sales amount that would have been obtained by directly debiting the Sales account. This is true because the Sales Returns and Allowances account is a *contra* account to gross sales on the income statement. Assume that gross sales were $1,000 and sales returns and allowances were $50. This information would be presented on the income statement as:

Gross sales	$1,000
Less: Sales returns	
and allowances	50
Net sales	$ 950

Therefore, under either method the same net sales amount is obtained, but, by utilizing the Sales Returns and Allowances account, valuable information is preserved.

Sales Discount

A sales discount is primarily a marketing device used either to maintain parity with competition or, by offering better credit terms, to gain a marketing advantage over the competitors. Whether the firm's sales discount policy is effective can be evaluated by reviewing the number of sales discounts that have been taken by the customers.

Credit terms may be stated as 2/10, n/30, which means 2% discount if paid in 10 days from the date of delivery or the total amount in 30 days. The terms offered vary from firm to firm, but the accounting is the same. Even though the firms expect everyone to take advantage of the discount, the entry is not made on the basis of this expectation. Therefore, the entry to record a sale of $500 made on credit terms of 2/10, n/30 would be:

Accounts Receivable	500	
Sales		500

If the customer takes advantage of the discount and remits payment within 10 days, the entry is:

Cash	490	
Sales Discount	10	
Accounts Receivable		500

Note that the Sales account was not debited. Rather, a separate account, Sales Discount, was created so that the amount of the discounts for the period could be determined. This is the same rationale used for the creation of the Sales Returns and Allowances account. The Sales Discount account is also reflected on the income statement as a *contra* account to gross sales.

If the customer chooses not to save $10 by paying within 10 days he must still pay within 30 days. The entry for the collection of the account after ten days is the typical collection entry:

Cash	500	
Accounts Receivable		500

In this case, the customer is paying $10 or 2% of the purchase price for the use of funds for 20 days. This is an extremely high interest rate on a yearly basis. Two percent for 20 days is a yearly interest rate of 36%. The calculation is:

$$\frac{\text{Discount rate}}{\substack{\text{Length of time from} \\ \text{end of discount period to} \\ \text{when the amount is paid}}} \times 360^a = \text{Yearly rate}$$

$$\frac{2\%}{20 \text{ days}} \times 360 = 36\%$$

[a]For computational purposes, a common business practice is the use of a 360-day year.

A rate of 36% is considerably higher than the interest rates charged by banks. Therefore, management should borrow funds to take advantage of the discount. If the discount is not taken, payment should not be made until the due date because paying earlier increases the effective interest rate.

If few customers are taking advantage of the discount policy either they are not considering the effective cost of lost discounts or the discount is so low that to lose the discount makes little difference. The absence of a high percentage of customers taking the discount should be investigated to determine the cause. If the answer is a low discount rate, an increase of the discount may promote a significant increase in sales.

Uncollectible Accounts Receivable

Thus far, only those receivables that will be collected have been considered. Firms do not extend credit unless they can reasonably expect payment. However, even the best efforts of management cannot guarantee the collection of all receivables. In addition, an objective of management is to increase net income. The collection of accounts receivable is an important goal, but it is only part of increasing net income. Stringent credit policies will not only reduce the number of uncollectible accounts but will also reduce sales. The net effect could be a reduction in net income.

Because some receivables will not be collected, there must be accounting procedures to deal with these exceptions. Receivables that will not be collected have no value and cannot be shown as assets. In addition, since the receivable was not collected the firm, in effect, did not have revenue; however, a sale was made and the merchandise delivered. Thus the accounting procedure is to show the uncollectible amount as an expense rather than a reduction of sales. This is a natural extension of the matching concept. The revenues earned and the expenses incurred in a period are related. The fact that certain sales made in this period are not going to be collected constitutes an expense of the period in which the sales were made.

Assume that the ABC Company determined that a $2,000 receivable from J. Smith will not be collected. The bad debt could be recorded by the following entry:

Bad Debt Expense	2,000	
Accounts Receivable — J. Smith		2,000

By debiting the Bad Debt Expense account, the firm has reduced the net income by the amount of the original sale. This entry negates the effect of the original sale and provides management, through the use of the Bad Debt account, information relative to the effectiveness of the credit department.

The credit to the Accounts Receivable of J. Smith is possible because it is known that J. Smith will not pay. The use of this method is known as the *direct write-off* of accounts receivable. While this method is not generally recommended it is often used in those firms with relatively few credit sales and by small or professional businesses.

Allowance for Uncollectible Accounts. The preferred method of accounting for uncollectible accounts is through the use of an allowance account. At the end of the period, the collectibility of accounts receivable will not be known with certainty; however, the revenue has been reflected in the current period. The matching concept dictates that the expenses must be related to the revenues. The sales were made in one period so that the expenses must be estimated in the same manner for the same period. This will require an adjusting entry of the following type:

Bad Debt Expense	xxx	
Allowance for Uncollectible Accounts		xxx

Note that in the adjustment, an allowance or *contra* asset account is credited rather than the Accounts Receivable account. The reason for this is that, in making the adjustment, the firm is not specifying who is not going to pay, just that some of the receivables will not be collected. If the firm credited Accounts Receivable it would be necessary to remove certain accounts from the subsidiary accounts receivable ledger. The Accounts Receivable control account is the summation of the individual customers accounts. Therefore, a credit to the controlling account would require a credit to a subsidiary account. Since the identity of the individual who is not going to pay is unknown, the controlling account cannot be credited.

The estimate of the amount of bad debts usually will be based upon past experience, modified in accordance with current business conditions. Increased business activity and high employment should lead to low collection problems, while a declining economy and high unemployment would have the opposite effect.

The fact that the expense is an estimate necessarily means that personal judgment becomes a factor. The reasonableness of the expense is the important consideration. The following methods for determining the expense result in a specified amount; however, the expense actually incurred may not be the same as the calculated amount. Note that there are two major reasons for using the allowance method for uncollectible accounts:

1. Appropriate matching of revenues and expenses on the income statement.
2. Proper valuation of the receivables on the balance sheet.

Methods for Calculating Uncollectible Expense

Two alternative approaches are used in determining estimated uncollectible accounts. One method consists of adjusting the allowance account to a new balance equal to the estimated uncollectible portion of the existing accounts receivable. This method emphasizes that the receivables at the end of the year are the only items that have not been collected. Therefore, a review of those accounts is required before the proper balance in the allowance account can be determined. The balance in the allowance account becomes the focal point and the amount of expense to be charged is of secondary importance. Thus, this approach emphasizes the balance sheet rather than the income statement. The primary technique employed in this method is the use of an aging schedule for accounts receivable.

The second method requires that the adjusting entry be based on a percentage of the current year's net sales. The percentage of sales method emphasizes the expense side of the adjustment without consideration of the balance in the allowance account. Consequently, this method emphasizes the income statement rather than the balance sheet. If, over the course of time, an excessive balance develops in the allowance account, a reduction in the percentage of net sales is recommended.

Either of the preceding methods should lead to the desired result. That is, the expense incurred because of the failure to collect the receivable is related to the same period as the sale. As with any alternative accounting methods, the method selected should best represent the situation. In a given situation one method may be better than the other; consequently that method should be used. These two methods are explained as follows:

Percentage of Accounts Receivable.   As stated previously, the reason for the adjustment of the allowance is to value the receivables on the balance sheet and to relate expense for uncollectible receivables to the time period in which the sale was made. The easiest method is to find the amount of receivables from the ending balance that will not be collected. This is conceptually the easiest method, but unfortunately customers rarely inform the firm that they will not pay until they are bankrupt, or unable to pay. Even without the knowledge of the amount of the uncollectible accounts, the firm can still make an approximation.

There are two accepted methods for approximating the amount:

1. *Arbitrary percentage of the receivables balance.* A certain percentage of the receivables balance is assumed to be uncollectible. The percentage is said to be arbitrary, not because one individual specifies the percentage, but rather because only one percentage is used. This percentage, e.g., 5% or 10%, should be determined carefully, utilizing past information of the firm or industry averages. For example, if the receivables ending balance is $80,000 and it is expected that 3% will not be collected, the following entry will be made:

Bad Debt Expense	2,400	
Allowance for Uncollectible		
Accounts		2,400

Presumably, during the next accounting period $2,400 of accounts will not be collected from this year's balance of receivables. If the amount uncollected is not exactly $2,400, the adjusting entry next year will reflect this fact. In this method, the adjustment is made for the amount necessary to bring the balance in the allowance account to the specified percentage of the current year's ending receivables balance.

In the previous example, if the allowance account had a credit balance at the time of adjustment in the amount of $200, the adjustment would have been as follows:

Bad Debt Expense	2,200	
Allowance for Uncollectible		
Accounts		2,200

Thus the overestimate of uncollectibles from the previous year is offset by an undercharge to expense for the current year. This is not theoretically correct from the point of view of periodic reporting; however, the difference was the result of an error in a previous estimate. In accounting, errors in estimates are not considered to be justification for restating prior period financial statements.

2. *Aging the receivables account.* This method provides for grouping the receivables by the age of each account. The assumption is that the longer an account has been uncollected, the less likely that it will be collected. The fact that an account is past due suggests that the customer cannot or will not pay. This method is known as aging the receivables and is illustrated in the following example:

XYZ Company has a balance in the receivables account on December 31, 1974 of $80,000. The accounts are reviewed and placed in the time periods as indicated in the table. The percentages used represent the best estimate of the amount of the receivables in each time period that will not be collected.

Age (days)	Amount	×	Percentage	=	Uncollectible
1–45	$50,000		1%		$ 500
46–90	15,000		3%		450
91–150	10,000		10%		1,000
Over 151	5,000		50%		2,500
	$80,000				$4,450

This analysis gives management useful information regarding the percentage of the receivables past due. The aging schedule also vividly presents the probability of collecting the past due accounts and the amounts in each category.

An important point to be noted is that the time intervals selected should be representative of the firm's credit terms. If the credit terms are net 60 days, there is no reason to set up a 0–30 day interval. Therefore, the time intervals selected should reflect the credit terms of the firm.

In the example the aging of the accounts indicated a need for a balance in the allowance account of $4,450. Before an adjusting entry can be made, the current balance in the account must be known.

The total amount of the provision, in this case $4,450, is presumed to be required for accounts which will be uncollectible in subsequent periods. However, as has been noted, this calculation is only an estimate and may not be accurate; consequently, at the end of any year the allowance account may have a balance. The amount of the balance should be small, since presumably the previous entries reflected the best estimates of uncollectible receivables. If there is a $700 credit balance in the allowance account the adjusting entry would be:

Allowance for Uncollectible Accounts

	700 Current balance @ 12/31
	3,750 Adjustment
	4,450 Required balance

Bad Debt Expense	3,750	
Allowance for Uncollectible Accounts		3,750

This is the same procedure used in taking a percentage of the receivables to arrive at the desired provision. If the allowance account had a debit balance of $300, the adjusting entry would be made in the amount of $4,750 (4,450 + 300) in order to create the desired credit balance of $4,450.

Estimating the Uncollectibles as a Percentage of Net Sales

An alternative approach of providing for uncollectible accounts consists of computing the charge to bad debt expense based on a percentage of the net sales for the year. This method emphasizes the expense involved rather than concern for a proper balance in the allowance account. This places the emphasis on the income statement rather than the balance sheet.

Even though this procedure is called a percentage of net sales, this is somewhat of a misnomer. Theoretically it should be a percentage of *net credit sales*, not net sales, because included in net sales are cash sales, which cannot be bad debts. However, if the cash sales are not known, or if they are a small portion of net sales, the amount of the bad debt expense will not be materially effected. Therefore, the common solution is to use an adjusted rate which reflects the inclusion of cash sales.

As an example of this method, assume that the bad debt expense has consistently averaged 1½% of net credit sales. At the end of the current year and before adjusting entries, the following information is known concerning XYZ Company:

Total sales		$2,400,000
Less: Cash sales		− 310,000
Credit sales		$2,090,000
Less: Sales returns and allowances	$20,000	
Sales discount	70,000	90,000
Net credit sales		$2,000,000

The net credit sales for the current year amount to $2,000,000 and 1½% of this amount is $30,000. Therefore, the entry for the bad debt expense is:

Bad Debt Expense	30,000	
Allowance for Uncollectible Accounts		30,000

Note that any balance existing in the allowance account is not considered in making the preceding entry.

Write-Off of an Account Receivable

Whenever an account receivable from a customer is determined to be uncollectible, it is no longer an asset and must be *written off* the books. In writing the account off the books the customer's account is reduced to zero. The allowance account must also be reduced by the amount of the uncollectible account.

EXAMPLE. Assume that the records of JKL Company indicate balances in Accounts Receivable of $75,000 and in Allowance for Uncollectible Accounts of $8,000. JKL learns that a customer, George Baker, is bankrupt and his account receivable of $850 will never be collected. The following entry should be made:

Allowance for Uncollectible Accounts	850	
Accounts Receivable		850

After the entry writing off George Baker's account, the records will appear as follows:

Accounts Receivable

1974		1975	
Dec. 31	75,000	Jan. 2	850
Balance	74,150		

Allowance for Uncollectible Accounts

1975		1974	
Jan. 2	850	Dec. 31	8,000
		Balance	7,150

The illustration presented demonstrates that the entry has no effect on the balance sheet, because an allowance was already provided for accounts of this type.

	Dec. 31, 1974	Jan. 2 1975
Accounts receivable	$75,000	$74,150
Less: Allowance for uncollectible accounts	8,000	7,150
Net value of receivables	$67,000	$67,000

Consequently, there is no expense incurred in writing off an account. The expense resulting from the failure to collect the receivable was related to the period in which the sale was made. This was accomplished by an adjusting entry which was made at the end of 1974.

Recovery of Accounts Previously Written Off

The accounts that were written off were the result of attempts at collection which proved futile. Usually accounts are written off when they are long past due, but sometimes dramatic events, such as bankruptcy, provide the impetus for the write-off.

Occasionally, accounts which have been previously written off will be collected in full or in part. Recovery of bad debts is evidence that the write-off was in error. Therefore, the entry made in writing off the account should be reversed. By reversing the entry, the account is taken from the bad credit risk file and put back into the active file. This provides management with a record of the charges and credits to the individual customer.

The entry to record the recovery of the $850 account of George Baker previously written off is:

Accounts Receivable — George Baker	850	
Allowance for Uncollectible Accounts		850

This entry puts the account back on the books and then an entry is made for the receipt of cash.

Cash	850	
Accounts Receivable — George Baker		850

A partial recovery will be recorded by reversing the write-off for the amount of the payment.

Balance Sheet Presentation — Accounts Receivable

The controlling account balance is listed on the balance sheet, even though accounts receivable is a summation of the individual account balances owed to a firm. The reasons for this are obvious: (a) a listing of individual balances would require considerable space on the balance sheet, and (b) the reader of the financial statements would not have any more useful information. Information about the customers of the firm might be interesting, but this curiosity could lead the reader to overlook the importance of total receivables.

The accounts receivable controlling balance is presented on the balance sheet in the Current Asset section. In addition, the balance in the allowance account is deducted from the receivables. This permits the reader to see the expected realizable value of the receivables. The typical presentation is as follows:

Accounts receivable	$95,000
Less: Allowance for uncollectible accounts	−12,000
Net accounts receivable	$83,000

NOTES RECEIVABLE

The majority of credit sales are on open-book accounts (Accounts Receivable). The customer submits an order and the seller ships the merchandise. The seller keeps a record of the payment that is due but does not require a signed statement from the customer acknowledging the debt; however, the seller may require a signed note rather than selling on an open-book account.

Two examples are:

1. Customer does not have the money now but will have it in six months or a year. An interest rate is attached to the note for the delay in receiving the amount due.
2. Customer originally has an open-book account but when payment is due, the seller accepts a note with an interest rate attached in lieu of the cash payment.

A simple signed note (commonly called a promissory note) from a customer does not give rise to any special rights for recovery on the part of the seller. However, if the invoice is not paid, the note constitutes persuasive legal evidence of the validity of the purchaser's obligation. The entry to record the receipt of a $1,000, one-year, 7% note in exchange for an accounts receivable is as follows:

Notes Receivable	1,000	
Accounts Receivable		1,000

There is no recognition of any interest at this time. Interest is earned over the life of the note. Thus when cash is received at maturity the entry is:

Cash	1,070	
Notes Receivable		1,000
Interest Income		70

The preceding entry presumes that the note will be collected. If the note is not paid at maturity it is called a *dishonored note*. Immediately after the dishonor, an entry should be made transferring the amount due on the note receivable (principal plus interest) to an account receivable from the debtor.

The amount should be transferred from Notes Receivable to Accounts Receivable because a past due note is not a note for accounting purposes. The claim for payment is once again an open-book item, although the firm does have the signed promissory note as evidence of the indebtedness.

Assume that in the prior example the note was not paid at maturity. The entry to record the dishonored note is:

Accounts Receivable	1,070	
Notes Receivable		1,000
Interest Income		70

Note that the interest earned becomes part of the accounts receivable. The interest earned, while holding the note, is as valid a claim when the note is due as the face amount of the note.

Discounting a Note Receivable

Most firms are not in business to finance sales to customers. By accepting a note, the firm becomes a finance company and must wait until the maturity date of the note to collect payment. The financing of the other activities of the business is made more difficult and, consequently, the firm may find it necessary to sell the note to the bank. This is called *discounting* a note. The holder of the note signs his name on the note and gives it to the bank. The bank pays the holder of the note and accepts the responsibility of holding the note until maturity. The bank expects to collect from the maker of the note at maturity. However, if the note is not paid, the bank can assert a secondary claim against the original holder of the note for payment.

The endorser, or original holder of the note, has a contingent liability to pay the note if the maker fails to make the payment. A *contingent liability* may be regarded as a potential liability, i.e., the events necessary to make the obligation a certainty have not occurred. If the maker fails to pay at maturity the endorser has a liability; however, if the maker pays the note as expected, the endorser has no liability.

The existence of a contingent liability should be disclosed on the balance sheet; however, it is not classified as a liability. Rather, the contingency is disclosed by means of a footnote wherein the particular circumstances are set forth.

Computing the Proceeds

The amount that the endorser receives from discounting a note is called the *proceeds*. The following are the steps used in determining the proceeds:

1. Calculate the maturity value of the note. This is normally the face value of the note plus interest.
2. Determine the length of time that the bank will hold the note. Count the number of days from the day of discounting to the maturity date. *Count either the first day or the last day but not both.*
3. Compute the discount by the following formula:

$$\begin{array}{c}\text{Maturity} \\ \text{value} \\ \text{of note}\end{array} \times \begin{array}{c}\text{Discount rate} \\ \text{(interest rate} \\ \text{bank charges)}\end{array} \times \dfrac{\begin{array}{c}\text{Number of days} \\ \text{the bank will} \\ \text{hold the note}\end{array}}{360} = \text{Discount}$$

4. Subtract the discount, computed in 3, from the maturity value of the note, determined in 1, and the result is the proceeds.

To illustrate this procedure, assume that on May 1, Tom Madden received a 60-day, 7% note for $6,000 from James Flynn. The note matures June 30 (31 days in May and 29 in June). On May 30, Madden discounts the note at a bank which charges a discount rate of 6% per year. The computations are as follows:

Face value of note	$6,000.00
Interest to maturity (6,000 × 7/100 × 60/360)	70.00
Maturity value	$6,070.00

$$6{,}070 \times {}^{6}/_{100} \times {}^{30}/_{360} = \$30.35 \text{ Discount}$$

Maturity value of note	$6,070.00
Less: Discount	30.35
Proceeds of note	$6,039.65

The entry to record the discounting of the note is as follows:

Cash	6,039.65	
Notes Receivable		6,000.00
Interest Income		35.00
Financing Income		4.65

An alternative procedure is to credit Notes Receivable Discounted instead of Notes Receivable. When this account is used, it is presented as a *contra* account to Notes Receivable on the balance sheet. When payment is made by the maker of the note to the bank, an entry must be made debiting Notes Receivable Discounted and crediting Notes Receivable.

The interest income is the amount of the interest that Madden had earned on the note up to May 30 ($6,000 \times ^7/_{100} \times ^{30}/_{360} = \$35$). This amount must be calculated in order to determine the book value of the note, which is the value of the note to Madden on the day it is discounted. Madden discounted a note with a book value of $6,035 and received $6,039.65. The increase of $4.65 resulted from a favorable financing arrangement and is properly recorded as Financing Income.

The financing income resulted from the difference between the interest rate on the note and the discount rate charged by the bank (7% versus 6%). If these rates had been reversed, Madden would have received less than book value and would have incurred a financing expense.

Balance Sheet Presentation — Notes Receivable

The notes receivable are presented at their face value. The interest that has been earned is listed as interest receivable. Both of these items are presented in the Current Asset section of the balance sheet unless the note will not be collected within the current accounting period. In that case, they are long-term assets.

ACCOUNTS RECEIVABLE SYSTEM

The expansion of credit has increased the importance of maintaining adequate records for accounts receivable. For each account receivable, a firm must maintain a continuous record of amounts charged to the customer, the amounts credited as the result of payments, allowances made, and the balance due.

There are two essential record keeping steps involved in each account receivable.

1. An account must be established for each customer.
2. Records must be maintained for all transactions affecting each account and the aggregate accounts receivable.

The design of a system for accounts receivable involves considerable paperwork; however, reasonable speed and accuracy must be provided by the system so that:

1. Good customer relations will be promoted and maintained.
2. Collection activities will be facilitated to insure rapid cash inflow and minimize bad debt losses.
3. Internal control will be effective and any fraud attempted through manipulation of accounts made more difficult.

A properly designed accounts receivable system will generate information that will satisfy the needs not only of the credit department but also other departments. Sales departments, marketing departments, and others benefit from the information generated by a well-conceived accounts receivable system. In addition, a properly designed accounts receivable system is a prerequisite for insuring the integrity of the accounts receivable information for external users.

TYPES OF SYSTEMS

Manual System. In a manual accounts receivable system, invoices and credit memorandums are sorted daily into serial number order, entered in the sales journal, and filed in numerical order. Cash receipts are sorted daily and entered in the cash receipts journal. Additional information regarding customers' payments is filed in the order in which journalized.

These journals are posted to the subsidiary accounts receivable ledger at regular intervals. Postings should be made on a daily basis for good cash control. The journals will be totaled at the end of the month and posted to the general ledger accounts receivable control account.

At the end of the month, after the posting is completed, a trial balance of customers' accounts is prepared. This total must agree with the balance in the control account.

The weakness of this system is that errors are detected only at the end of the month. If the accounts are numerous, the balancing should be completed on a more frequent basis. As is evident, this system can only be used when the volume of accounts receivable is small. Too many receivables would involve a substantial labor cost, probably exceeding the potential benefits.

Semiautomatic Method. The use of semiautomatic equipment may be sufficient to overcome the weakness of the manual system. For firms without substantial accounts receivable, small office machinery, such as typewriters, adding machines, and simple posting machines, may be adequate. Because of the lack of a clear distinction between semiautomatic and more mechanized methods, further discussion will be included in the following section, which illustrates a computerized accounts receivable system.

Automated System. Jobs that are repetitive can be easily programmed and assigned to electronic computers. The processing of accounts receivable is one of the most common functions performed through a computer operation. A large number of companies have attained a measure of automation in the processing of accounts receivable. The computer is useful in this area because the sale transactions and the collection of the receivables are repeated on a daily basis. Thus a tremendous amount of paperwork which would be generated by these transactions is eliminated through the use of a computer.

The following example is used to demonstrate the need and advantages of using the computer in processing accounts receivable.

Knobtons, Inc. is a large wholesaler with several branch offices averaging 5,000 billings per month. They used cycle billing so that all bills are not sent out on one day. (*Cycle billing* is a procedure whereby bills are mailed on an alphabetical basis.) This procedure spreads the workload throughout the month. In addition, the cash inflow is spread over the month so that cash management is facilitated.

Invoices are prepared manually at the branch offices and copies are forwarded to corporate headquarters. The invoices are put into punched card form and processed by the computer. In this way, a centralized accounts receivable file is prepared in the computer. This means that Knobtons has computer access to corporatewide sales figures, type and name of customer, amount of sale, and related information. These data then serve as the basis for management reports and decisions. The same information could be attained in a manual system, but because of the volume of work no manual method would be practical.

The customers mail their checks to bank lockboxes where they are processed each day by the local banks. The banks mail photocopies of the checks to corporate headquarters. In this manner, Knobtons receives an immediate deposit record of each remittance.

Once remittance information is received in the data processing center, a cash receipts card for each check (customer identification, remittance amount, invoice numbers if indicated, etc.) is keypunched. These cards serve as the input to the computer. A computer program would be designed to match the remittances to open items on the customers' accounts receivable file. Therefore, at any time, a listing of the accounts receivable, the age of each account, and the total uncollected can be obtained.

As the volume of work increases, the necessity for more automated equipment becomes more apparent. Manual methods require too much labor cost, and the length of time involved to obtain the desired informa-

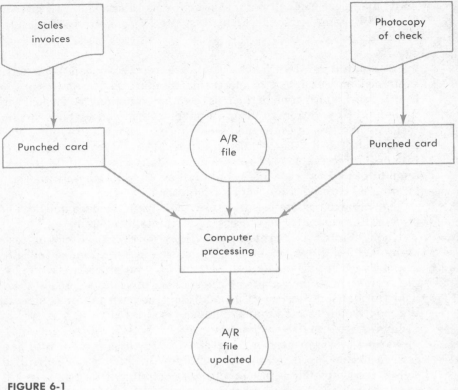

FIGURE 6-1

Simplified automated accounts receivable system for Knobtons, Inc.

tion is such that the information is often useless when received by management.

The accounts receivable system employed depends upon the firm; however, some of the advantages that can be expected if a computer is used are:

1. Savings in the cost of processing the receivables.
2. Increased accuracy because the human element is minimized after the information has been keypunched.
3. Increased speed relative to mailing customer statements.
4. Accelerated flow of remittance information to the credit department.
5. Aging schedules prepared so that management can limit their examination of receivables to those accounts.
6. Controls over receivables become more effective.

SUMMARY

The major type of receivable is accounts receivable, which arises from the sale of goods and/or services that the firm is in business to sell. Accounts receivable must be given special consideration by management because of volume, credit policies, and impact on profitability, i.e., meeting competition and bad debts. In addition, the volume of paperwork and the potential for defalcations associated with accounts receivable are such that a properly designed system is an absolute necessity. Many firms have found that a manual system is inadequate for their needs and an automated system necessary and economical.

QUESTIONS

6-1. What are the essential elements of a receivable?

6-2. Why is the return of merchandise not recorded as a debit to the sales account?

6-3. What is the effective cost of missing the discount if the terms are 5/20, n/60? 4/40, n/90?

6-4. Why should estimated bad debts be recognized prior to the time they actually prove to be uncollectible?

6-5. There are two approaches for estimating the adjustment for uncollectible accounts. What are these approaches and explain the difference?

6-6. Company B has accounts receivable of $274,650 and a balance in the Allowance for Uncollectible Accounts of $9,750. Management decides that an account in the amount of $3,600 is uncollectible and should be written off the books. What is the effect of the write-off on the current income statement? What is the effect on total current assets? Why?

6-7. The ABC Company computes its bad debts as 1% of net credit sales. Credit sales have remained constant for several years; however, the balance in the allowance account has continued to increase. What suggestions would you make to the company?

6-8. Determine the maturity date of the following notes:
(a) 30-day note dated May 18
(b) 60-day note dated July 1
(c) 90-day note dated April 14

6-9. What is discounting a note? Describe the procedure used in computing the proceeds of a discounted note.

6-10. How is a contingent liability shown on the balance sheet?

6-11. Systems for accounts receivable generate information for other departments in addition to the accounting department. Describe some information that would benefit the marketing department?

6-12. Most firms make some use of automated accounts receivable systems. Why?

PROBLEMS

Procedural Problems

6-1. The following are the balances in selected accounts of the Smith Company as of December 31, 1975. Prior to closing the temporary accounts, management decides to write off an account in the amount of $1,800 of the Roebuck Company. (This was a credit sale of a previous year.)

Accounts receivable	$ 61,800
Allowance for uncollectible accounts	1,500 Dr
Sales	726,000
Sales returns and allowances	16,000
Sales discounts	10,000

REQUIRED

(a) Prepare a journal entry to record the write-off of Roebuck Company. Also prepare an entry to record the bad debt expense for the period assuming the estimated expense is 1½% of net sales.

(b) Prepare the entry to record the estimated expense for the period if the allowance is to be adjusted to 6% of outstanding receivables.

6-2. The following were selected from among the transactions completed by Sunshine Sales Company during the current year:

Jan. 3 Sold merchandise on credit to Harold Hall, $525. (Terms of all credit sales are 2/10, n/30.)

Jan. 20 Sold merchandise on credit to John Thomas, $840.

Jan. 28 Collected amount due on account from John Thomas.

Feb. 2 Collected amount due on account from Harold Hall.

Mar. 24 Sold merchandise on credit to Tom Smith, $1,820.

Apr. 1 Tom Smith returned merchandise of $300.

Apr. 2 Collected the amount due on account from Tom Smith.

REQUIRED. Journal entries for the preceding transactions.

6-3. The Scott Appliance Company ages its accounts receivable and applies the following percentages to determine the correct balance for the Allowance for Uncollectible Accounts:

1–30 days	1%	91 days to 6 months	30%
31–60 days	5	Over 6 months	100
61–90 days	12		

The aging process resulted in the following breakdown of accounts receivable:

1–30 days	$34,250
31–60 days	9,530
61–90 days	5,360
91 days to 6 months	2,620
Over 6 months	940
	$52,700

The current balance of Allowance for Uncollectible Accounts is $1,237. Prior to preparing its year-end adjusting entries, the company writes off all accounts over six months past due.

REQUIRED
(a) Completed aging schedule.
(b) Journal entry to write off accounts over six months past due.
(c) Adjusting journal entry for uncollectible accounts.

6-4. The following items are all related to uncollectible accounts of Burns Company and were completed during the current fiscal year ended December 31.

Jan. 20   Wrote off the $396 balance owed by Mitchell Corporation, a bankrupt company.

Feb. 28   Received $230 from Southland Corporation in full payment of its account which had been written off as uncollectible in the previous year.

June 7   Wrote off the following accounts as uncollectible: Frank Smith, $137; Roger Painter, $216.

Sept. 13   Received $560 as settlement of the $800 owed by Bacon & Co., a bankrupt. The remainder was written off as uncollectible.

Dec. 31   An analysis of the $55,400 accounts receivable indicates that $1,475 will be uncollectible. On January 1, the allowance account had a balance of $870.

REQUIRED
(a) Prepare the journal entries to record the preceding transactions and adjustments.
(b) Determine the expected realizable value of the accounts receivable as of December 31.
(c) Assuming that the direct write-off method was used, record the transactions of June 7.

6-5. John's Bargain Basement received a note from a customer with the following terms:

Face value of note	$10,000
Date of note	6/15/74
Life of note	60 days
Interest rate	6%
Note discounted	7/30/74
Discount rate	7%

REQUIRED.   Calculate:
(a) The length of time the company held the note.
(b) Maturity date of note.
(c) Maturity value of note.
(d) Dollar amount of the discount.
(e) Entry to record the receipt of the proceeds at the date of discount.

6-6. The Mattey Company completed the following transactions with respect to receivables:

Apr. 1   Received a $5,000, 90-day, 6% note from J. Jones for the sale of merchandise to him.

May 1   Received a $7,200, 60-day, 5% note from George Bradley in settlement of his account.

May 31   Discounted the Jones note at the bank at 7%.

June 1   Received a $9,600, 180-day, 7% note from James Stratton in settlement of his account.

June 15   Received a $3,000, 90-day, 8% note from Charles Talbot in settlement of his account.

July 1   Received payment on the Bradley note.

July 1   Received notification from the bank that the Jones note was not paid at maturity.

July 15   Discounted the Talbot note at the bank at 7%.

July 31   Received payment from Jones on his dishonored note, including interest at 6% on the maturity value of the note from the maturity date.

REQUIRED

(a) Prepare the journal entries necessary to record the foregoing transactions.

(b) The fiscal year ends on July 31. Prepare any necessary adjusting entries.

Conceptual Problems

6-7. Joe's Tape Center, which had previously sold only for cash, under pressure from competition opened charge account privileges for preferred customers. At the end of the first year under the new credit plan, Joe desired to appraise the results of his change in policy. He wanted to know if the credit plan should be (a) continued as is, (b) rela. ed, (c) made more stringent, or (d) returned to the original cash sales basis.

Joe's customers are chiefly young adults of middle income. The store is located in an older neighborhood and has been in operation three years.

What kind of financial information do you think Joe needs in order to make a decision on the future credit policy? Are there other nonquantitative factors that might have a bearing on the decision?

6-8. During a review of accounts receivable, the president of your company asks why the current year's expense for bad debts is charged merely because some accounts may become uncollectible next year. He then said that he had read that financial statements should be based upon verifiable, objective evidence, and that it seemed to him to be much more objective to wait until individual accounts receivable were actually determined to be uncollectible before charging them to expense.

REQUIRED

(a) What is the justification for the allowance method as contrasted with the direct write-off method of accounting for bad debts?

(b) Describe the following two methods of estimating bad debts:
  (1) The percentage of sales method.
  (2) The aging method.
(c) Of what merit is your president's contention that the allowance method lacks the objectivity of the direct write-off method?

6-9. The Salmac Automobile Insurance Company has over 600,000 policyholders. The company mails semiannual premium notices. Most of the policyholders remit the correct amount of premium by the due date, and thus no accounts receivable are generated. However, for a variety of reasons, the company always has between 25,000 and 30,000 individual accounts outstanding. The particular policyholders involved are constantly changing so that an account receivable subsidiary ledger page probably would only contain the original entry and the payment.

REQUIRED. Keeping in mind that up-to-date records are especially important to this company because of policyholder claims for accidents:
(a) Describe a suitable semiautomated accounts receivable system.
(b) Describe a suitable automated accounts receivable system.
(c) Indicate which you consider most appropriate for this company, giving consideration to speed, accuracy, and costs.

6-10. The Smith–Jones Company has been in operation for several years. The firm has followed the policy of recording bad debt expense using the direct write-off method. The company is now giving consideration to changing to the allowance method and management wants to know the percentage to use. Therefore, information is requested about past write-offs, the actual percentage of accounts written off, and the fairness of matching expenses and revenues during the past four years. Assume that you have determined the following information:

Year	Sales	Accounts written off	1971	1972	1973	1974
				Year of sale of accounts written off		
1971	300,000	1,500	1,500			
1972	550,000	2,600	1,200	1,400		
1973	800,000	6,900	300	4,500	2,100	
1974	1,200 000	8,000		400	6,700	900

REQUIRED
(a) What percentage of sales have been uncollectible in the past?
(b) Suppose the percentage method had been used, utilizing the percentage calculated in 1, what effect would it have had on bad debt expense?
(c) What effect did the direct write-off method have on the proper matching of expense and revenue?

6-11. The Ace Trucking Company, in settlement of an accounts receivable, accepted on June 1, 1974 a $2,500, one-year, 7% note of Champion Distributing Company. On October 1, 1974, Ace discounted the note with a local banker at a discount rate of 6%. One of the younger staff members questioned this transaction. The staff member stated that discounting the note resulted in the firm losing the 7% interest. In addition, there was no concern about Champion not paying, so why did Ace discount the note?

REQUIRED

(a) Answer the staff member's question regarding the reason or reasons for discounting the note.

(b) What are the proceeds from the note on October 1?

(c) If financial statements were prepared on December 31, what recognition, if any, should Ace make of the fact that the note had been discounted?

6-12. You are reviewing the accounting records of the Honolulu Branch of the Midwestern Distributing Company. This branch has substantial annual sales, which are billed and collected locally. As a part of the review you find that the procedure for collections on account is as follows:

Mail is opened by the secretary to the credit manager and remittances are given to the credit manager for his review. The credit manager then places the remittances in a tray on the cashier's desk. At the daily deposit cut-off time, the cashier delivers the checks and cash on hand to the assistant credit manager who prepares remittance lists and makes up the bank deposit, which he also takes to the bank. The assistant credit manager also posts remittances to the accounts receivable ledger cards and verifies the cash discount allowable.

You also ascertain that the credit manager obtains approval from the executive office of Midwestern Distributing Company, located in Chicago, to write off uncollectible accounts, and that he has retained in his custody, as of the end of the fiscal year, some remittances that were received on various days during the previous month.

REQUIRED. What is wrong with the procedures now in effect?

7

INVENTORY

EDUCATIONAL OBJECTIVES

The material in this chapter is designed to achieve several educational objectives. These include:

1. An awareness of the importance of good inventory records for those businesses where inventory is a significant factor in their operations.
2. A knowledge of the difference between perpetual and periodic inventory systems and the circumstances under which each would be employed.
3. A knowledge of the alternative methods for valuing inventory and the effect of the choice of methods on both the income statement and balance sheet.
4. An understanding of the impact of errors in the ending inventory on both the income statement and balance sheet.

The term *inventory* represents merchandise held for sale in the normal course of business. In addition, material used in production is also considered to be inventory. Therefore, inventory is especially important for a manufacturing or retailing firm since these firms generate revenue when they sell the product or products that they are in business to sell.

Manufacturing and retailing firms must also be concerned about inventory because of the amount of money invested in this asset. For example, in a recent financial statement for a manufacturing firm, inventories were $14,039,576. Total current assets were $27,101,361; therefore, inventories represented almost 52% of the current assets. In addition, inventories represented more than 34% of the total assets. These percentages typify the significance of inventories in relation to assets for most manufacturing firms. Thus analysis of the current position of a firm should include some consideration of the nature of the inventory and how it is valued.

Inventories are not important to the same extent for firms other than manufacturing or retailing ones. Banks, service organizations, and other similar institutions normally sell services or products which do not need special packaging and handling, protection from deterioration, or products which are subject to changes in taste or fashion.

Objectives for Inventory

Managerial and investor interest in inventory is focused on three primary objectives:

1. Income determination. In a business which has inventory, the measurement of net income is directly influenced both by the cost assigned to the inventory and the valuation method used in determining ending inventory.
2. Maximizing liquidity–profitability. There should be enough inventory on hand to meet expected demand, but, at the same time, not so much inventory that liquidity (cash position of the firm) is impaired.
3. Adequate controls and safeguards for inventories. The amount of money and the volume of goods necessitate controls that will reasonably insure that goods will not be pilfered or lost; however, the controls are not a perfect assurance that the inventory is totally protected.

To attain these objectives, there must be useful (relevant) information regarding inventories. This information must include the cost of items, volume of items, location, etc. The system to account for inventories should be designed to accomplish these objectives. As an example of how some inventory information is actually collected, consider the following situation. When the customer makes a purchase in a department store the cashier normally removes a perforated tag from the article. At the end of the day, the tags, collected from the items sold, are sent to the accounting department.

The accounting department then uses these data to develop cost-of-sales information and to keep a constant check on the inventory level of all items. When the stock of any item drops below the reorder point, a purchase order can be placed. Another common practice is the placing of a punched card in major appliances or furniture. As the item is sold the card is removed, sent to the accounting department, and serves the same function as the perforated tag in the previous example.

Perpetual Versus Periodic Inventory Records

As can be seen, the amount of paperwork involved in keeping accurate daily records for all items in an inventory would be time-consuming and expensive. As a consequence, daily records are generally kept only for those items that represent a large part of the inventory. In this context, large part means either a significant percentage of the inventory or a large dollar cost. To lose a pack of gum through poor internal control is much different from losing an automobile. A firm wants to maintain tight control and more information concerning an automobile but would accept less information and looser control for gum.

Maintaining constant inventory records is called a *perpetual inventory system*. The absence of constant inventory records is called a *periodic inventory system*. In either case, an actual inventory count is made before the annual published financial statements are prepared.

The illustration of adjusting entries presented in Chapter 4 presumed a periodic inventory system. Thus the entry to record cost of goods sold was made only at the end of the period after making a physical count. Under a perpetual inventory system, an entry recording cost of goods sold is made each time sales are recorded. For example, assume merchandise costing $500 is sold on account for $700. The entries to record the transactions are:

To record the sale:		
Accounts Receivable	700	
Sales		700
To record the cost:		
Cost of Goods Sold	500	
Merchandise Inventory		500

By making an entry for the cost every time a sale is made, the accountant maintains a record of how much should be in the inventory. As a result, any difference between the actual count and the inventory records is isolated and can be investigated. In the periodic system, this procedure cannot be followed because the accountant does not know how much should be in the inventory. For a full comparison of the entries both for the perpetual and periodic systems see Figure 7-1.

To summarize:

1. *If the cost per unit is high and the volume low, a perpetual system should be employed, e.g., automobiles, mink coats, furniture, appliances, etc.*

2. *If the cost per unit is low and the volume high, a periodic system should be employed, e.g., gum, items in a grocery store, etc.*

Inventory Cost

The determination of the cost of inventory is more complex than recording the purchase price of the item. The reason for this is that things such as freight charges, discounts, and returns of defective merchandise must be included.

The cost incurred by the purchaser in transporting the goods purchased to the buyer's firm is called *freight-in*. Since the merchandise would not be available without the payment of freight, such charges are a valid cost of the inventory. If the goods are shipped F.O.B. (Free On Board) Destination, the freight cost should be minimal. However, if the shipment is F.O.B. Seller's Plant, the freight may represent a substantial percentage of the total cost of an item.

The purchase of merchandise should be recorded at net, i.e. reduced by the amount of any discount allowed by the seller. The rationale for this approach is twofold:

1. The failure to take a cash discount results from not having cash to pay the bill. This is not a cost of the item, but rather a financing cost, and it should be recorded as a discount lost and shown on the income statement as a financial expense under the caption Other Income and Expense.

2. Management should understand the importance and significance of missing cash discounts and thus be expected to take all possible discounts. This is the opposite of the situation with accounts receivable, wherein the sales were not recorded, net of discounts. A seller cannot ascertain whether all of the customers will take the discount; however, a buyer is only concerned with his firm, and consequently, he can record purchases net of the discounts.

After the goods have been recorded at the net amount, any merchandise returned as a result of a damaged condition on delivery must be recorded. This information is important to management in the evaluation of suppliers and the firm's own purchasing department.

In Chapters 3 and 4, all costs associated with inventory were recorded in the Merchandise Inventory account. This procedure was used to facilitate the explanation of the accounting cycle; however, there are differences in accounting procedures depending upon the inventory system used by the firm. In a perpetual system, all costs are included in the Merchandise Inventory account and the account is reduced each time a sale is made. In a periodic system, several accounts are used to record the cost of inventory and the various accounts are closed at the end of the fiscal period into the Merchandise Inventory account and Cost of Goods Sold.

If, in Figure 7-1, the payment for the purchase had not been made within the discount period, the entry would have been:

Accounts Payable	294	
Discounts Lost	6	
Cash		300

This emphasizes the fact that the failure to take the discount is a financing cost not an inventory cost. As was mentioned, the Discounts Lost account is shown on the income statement as a financing expense.

Physical Inventory and Cost of Goods Sold

Most firms take a physical inventory at the end of each year whether they are using perpetual or periodic inventory systems. In some cases the inventory count is made on a more frequent basis. This asset is of such importance that common sense dictates that, periodically, a count should be made to determine whether the quantity that is assumed to be in the inventory is really present.

Errors in inventory affect both the balance sheet and the income statement. The effect on the balance sheet results from the error in reporting the value of the inventory. This mistake, therefore, has an effect on any financial analysis involving current assets. Consequently, extreme care must be employed in calculating ending inventory for balance sheet purposes.

A proper ending value is also of considerable importance for income statement purposes because the ending inventory directly affects profit. If the ending inventory is understated, net income is understated. If ending inventory is overstated, net income is overstated. The reason for this is that the ending inventory is a subtraction from the total goods available for sale. The higher the ending inventory, the lower the cost of

Date	Explanation	Perpetual system		Periodic system	
June 10	Purchase of 100 units @ $3, terms 2/10, n/30	Merchandise Inventory Accounts Payable	294 294	Purchases Accounts Payable	294 294
June 10	Paid freight charges $40	Merchandise Inventory Cash	40 40	Freight-In Cash	40 40
June 18	Paid supplier for goods purchased June 10	Accounts Payable Cash	294 294	Accounts Payable Cash	294 294
June 22	Sold 50 units @ $5	Cash Sales Cost of Goods Sold Merchandise Inventory	250 250 167 167	Cash Sales	250 250
June 27	Returned 10 units to supplier due to defective part	Receivables from Supplier Merchandise Inventory	33.40 33.40	Receivables from Supplier Purchase Returns and Allowances	33.40 33.40
June 30	Physical inventory count shows 40 units on hand			Cost of Goods Sold Purchase Returns and Allowances Merchandise Inventory Purchases Freight-In	167.00 33.40 133.60 294 40

FIGURE 7-1

goods sold and vice versa. Cost of goods sold, in turn, affects the gross margin. The full effect is illustrated by:

EXAMPLE I

Sales		$200,000
Cost of goods sold:		
Opening inventory 1/1/74	$ 40,000	
Purchases	150,000	
Total goods available for sale	$190,000	
Less: Ending inventory 12/31/74	60,000	
Cost of goods sold		$130,000
Gross margin		70 000
Operating expenses		40,000
Net income		$ 30,000

Assume it was discovered that $15,000 of inventory had been omitted in the inventory count. In this case, the income statement would have been:

EXAMPLE II

Sales		$200,000
Cost of goods sold:		
Opening inventory 1/1/74	$ 40,000	
Purchases	150,000	
Total goods available for sale	$190,000	
Less: Ending inventory 12/31/74	75,000	
Cost of goods sold		115,000
Gross margin		$ 85,000
Operating expenses		40,000
Net income		$ 45,000

The effect of the understatement of the inventory is the understatement of net income by $15,000. As an exercise, calculate the effect on the gross margin if the ending inventory in *Example I* included an overstatement of $25,000. (*Answer*: gross margin should have been $45,000 or was overstated by $25,000. The net income should have been $5,000 rather than $30,000, again an overstatement of $25,000.)

In summary:

1. *An understatement of ending inventory overstates cost of goods sold, understates gross margin, and understates net income.*
2. *An overstatement of ending inventory understates cost of goods sold overstates gross margin, and overstates net income.*

The ending inventory of this period is the succeeding period's opening inventory. Therefore, any error in the ending inventory produces an identical error in the opening inventory of the next period. As was shown, an error in the ending inventory affects net income in the same direction as the error in the inventory. The effect of an error in the opening inventory, however, is just the opposite. This is true because the overstatement of opening inventory overstates the cost of goods sold, which understates the gross margin and net income. Conversely, the understatement of opening inventory overstates net income.

Continuing with *Example II*, the omission of $15,000 of ending inventory in 1974 affects the income statement of 1975 in the following manner:

	Without considering the $15,000		Correct inventory value
Sales		$180,000	$180,000
Cost of goods sold			
Opening inventory			
1/1/75	$ 60,000		$ 75,000
Purchases	100,000		100,000
Total goods			
available for sale	$160,000		$175,000
Less ending			
inventory			
12/31/75	30,000		30,000
Cost of goods sold		130,000	145,000
Gross margin		$ 50,000	$ 35,000
Operating expenses		30,000	30,000
Net income		$ 20,000	$ 5,000

From the example, note that the understatement of the ending inventory on 12/31/74 understated the net income of 1974 by $15,000 and overstated the net income of 1975 by $15,000.

VALUATION OF ENDING INVENTORY

The physical count of inventory reveals only the number of items on hand, not the valuation of those items. The valuation can only be determined when the cost of the item is associated with the items. However, if there are 5,000 different products and 30,000 boxes or items counted, associating cost with particular items is extremely difficult. Consequently the specific identification of cost with an item in inventory valuation is used only on high cost-low volume items. For the majority of other products, some other inventory valuation method must be employed.

Inventory Valuation Methods

First-In, First-Out (FIFO). This valuation method is compatible with the flow of goods; however, inventory valuation methods are concerned with the flow of cost into cost of goods sold rather than with the flow of goods. Inventory, generally, must be sold on a first-in, first-out basis to prevent deterioration of the products. Over the long run, most products will deteriorate and, as a result, reshuffling of stock is a necessary practice in most firms. Supermarkets are a prime example since they must follow this practice or lose an enormous amount of inventory through deterioration. FIFO can be illustrated as follows:

FIFO Inventory Method

Beginning inventory	7 units @ $ 5 =	$ 35
First purchase	6 units @ $ 8 =	48
Second purchase	14 units @ $ 9 =	126
Third purchase	10 units @ $10 =	100
Total cost of goods available for sale		$309
Ending inventory	11 units 10 units @ $10 =	$100
	1 unit @ $ 9 =	9
		$109

In the example, 26 units were sold. The cost of these units represents the cost of goods sold and totals $200 (7 @ $5 = $35; 6 @ $8 = $48; 13 @ $9 = $117). Thus the cost of goods sold is composed of those units in the opening inventory plus the initial purchases in the period. Note that the ending inventory is composed of units from the most recent purchases.

Last-In, First-Out (LIFO). The argument in favor of LIFO is that this method does a better job of matching current costs with current revenues. This method does not depend upon actual sale on a LIFO basis, but rather it is based upon an assumption as to the flow of costs. With the LIFO method, the cost of the most recent purchases is transferred to cost of goods sold. Thus, the cost of the goods purchased in the period is matched with the revenue generated in the period. In contrast, FIFO transfers to costs of goods sold the opening inventory which was purchased in a prior period plus the earliest purchases of the current period. Consequently, an argument in favor of LIFO is that FIFO does not achieve a proper matching of current period revenues and expenses.

To illustrate the LIFO method in contrast with FIFO, the previous example is used:

LIFO Inventory Method

Beginning inventory	7 units @ $ 5 =	$ 35
First purchase	6 units @ $ 8 =	48
Second purchase	14 units @ $ 9 =	126
Third purchase	10 units @ $10 =	100
Total cost of goods available for sale		$309
Ending inventory	11 units 7 units @ $5 =	$35
	4 units @ $8 =	32
		$67

Notice the difference in the ending inventory using these two valuation methods (FIFO $109, LIFO $67). The cost of goods sold under LIFO consists of the most recently purchased units and totals $242 (10 @ $10 = $100; 14 @ $9 = $126; 2 @ $8 = $16). The difference of $42 is accounted for solely on the basis of a different valuation method. The total cost of goods available for sale is the same in both cases ($309) as well as the items that are in the inventory. Therefore the reduction in ending inventory values increases the cost of goods sold.

Arguments other than matching may influence the selection of the valuation method. In a period of rising prices, LIFO will give a lower inventory value and, consequently, a higher cost of goods sold and a lower net income. Lower net income means lower taxes and since lower taxes improves the liquidity position of the firm, the LIFO inventory method may be selected in periods of rising prices for tax reasons rather than for accounting reasons.

There are some people who argue that the tax aspect is the primary reason for the use of LIFO. Their reasoning is that the shift to LIFO has been brought about by rising prices and the Internal Revenue Service requirement that in order for LIFO to be used for tax purposes it must be used for accounting purposes.

The major argument in favor of LIFO—matching—is an income statement emphasis. When the two methods are viewed in terms of balance sheet presentation a different argument arises. In periods of continually rising prices, the LIFO method will use the oldest costs and hence will result in the lowest inventory value. Therefore LIFO may match revenue and expense better than FIFO, but it will present a distorted balance sheet valuation.

In summary, FIFO is advocated because the goods are sold on that basis; LIFO is advocated because this method provides a better matching of expenses and revenues. In addition, in a period of continually rising prices, LIFO will result in a lower profit, lower inventory value, and less taxes paid than FIFO. In a period of continually falling prices, the reverse is true.

Weighted Average Method. This method is based upon the assumption that, because the items on hand have the same physical characteristics as the items sold, an average cost per item may be used. Therefore the extreme values generated by LIFO and FIFO are avoided. Weighted average is a compromise method which has merit in a market typified by fluctuating prices; however, there is little theoretical justification for the method. As was indicated, inventory must be sold on a first-in, first-out basis to protect against deterioration. In addition, the weighted average flow of costs does not properly match the revenues generated in the period.

Mathematically, this method is the average cost obtained by multiplying the price per unit by the number of units purchased at those prices. Utilizing the same example as for FIFO and LIFO:

Weighted Average Inventory Method

Beginning inventory	7 units @ $ 5 =	$ 35
First purchase	6 units @ $ 8 =	48
Second purchase	14 units @ $ 9 =	126
Third purchase	10 units @ $10 =	100
Total cost of goods available for sale	37 units	$309

$$\text{Cost per unit} = \frac{\text{Total amount of cost}}{\text{Total number of units}} = \frac{\$309}{37} = \$8.35$$

Therefore, ending inventory is 11 units @ $8.35 or $91.85, which is approximately the average of FIFO and LIFO. The cost of goods sold is 26 units @ $8.35 or $217.15.

Other Inventory Valuation Procedures

There are methods of inventory valuation other than those just discussed. Among them are lower of cost or market, gross profit, and retail inventory. *Lower of cost or market.* This method is a departure from valuing inventory on the basis of historical cost. This departure is necessary when there is evidence that the value of the inventory has dropped below cost and there are indications that the fluctuation is permanent. The amount of decrease in value is removed from the inventory account and shown either as a separate expense item if the amount is material, or as an increase in the cost of goods sold on the income statement. This procedure results in a conservative balance sheet value for inventory. Lower of cost or market may be used in conjunction with any inventory valuation method when the current market value of the items is less than historical cost; however, for income tax purposes, lower of cost or market may not be used in conjunction with the LIFO inventory method.

Gross profit method. There are times when inventory is not available to be counted, e.g., fire loss. An acceptable method for approximating the inventory under these circumstances is to apply the gross profit margin method. Consequently, this method is not used for financial statement purposes under normal circumstances.

The calculation involves applying the gross profit margin to sales in order to determine the cost of goods sold. Once cost of goods sold has been determined, the ending inventory can be derived. For example, assume that the gross profit over the past four years averaged 25%, the ending inventory could be determined as follows:

Sales		$100,000	100%	
Beginning inventory	$ 40,000			
Purchases	90,000			
Total goods available for sale	$130,000			
Ending inventory	X	*55,000*		
Cost of goods sold	Y	Y	75%	*75,000*
Gross profit		Z	25%	*25,000*

The unknown Z is determined by taking 25% of sales or $25,000. If the gross profit is $25,000, then the cost of goods sold must be $75,000. Once the cost of goods sold is determined, the ending inventory is the difference between goods available for sale and cost of goods sold. In this example, the ending inventory is $55,000.

The gross profit method is best used when the gross margin is stable. If the margin fluctuates widely it is difficult to justify the use of any of the margins of the previous years. The probable solution to the problem of fluctuations is to use an average of the gross margins.

Retail inventory method. This method is used widely by those firms that have large amounts of inventory, all of which is displayed at retail prices, e.g., department stores. When inventory is taken in a department store, the items are counted and valued at their retail price. The retail price may then be reduced to cost on the basis of the applicable gross margin. This cost figure may be used for financial statement purposes. Since margins differ among like items and the number of items in the inventory is enormous, the costing of inventory for a department store would be a monumental task.

Rather than go through the laborious process of checking invoice prices and individual gross margins, the accountant assumes that the ending inventory is composed of an average of all the items that the firm sells. If this assumption is true, and if an average markup of all goods can be determined, then the entire inventory can be counted at retail and reduced to cost by the average markup. This procedure eliminates much of the mathematics involved in valuing the inventory and, if the assumption is true, the determination of the ending inventory should reasonably approximate the true inventory cost. If the assumption is not true, the retail method must be modified.

The retail inventory method can be illustrated by the following example, where the estimated cost of the inventory is determined as of June 30:

Retail Inventory Method

		Cost	Retail
June 1	Merchandise inventory	$180,000	$288,000
	Purchases during June	145,000	212,000
Total goods available for sale		$325,000	$500,000

$$\text{Ratio of cost to retail } \frac{325,000}{500,000} = 65\%$$

Sales during June			170,000
Ending inventory @ retail			$330,000
Ending inventory @ cost (330,000 × 0.65)			$214,500

Notice that the first step is the determination of the cost ratio. After this is calculated, the amount of ending inventory at retail is determined, and the cost ratio is applied in order to reduce the ending inventory at retail to the ending inventory at cost.

Balance Sheet Presentation

Inventory is presented at cost on the balance sheet in the Current Asset section. In the case of a manufacturing firm, three inventories may be shown – Raw Materials, Work-In-Process, and Finished Goods. For a retail firm only one inventory, Merchandise Inventory, is presented. The exception to the rule of reporting at cost is the use of the lower of cost or market. If market is used the cost of the inventory is shown parenthetically. In addition, the method of inventory valuation should be disclosed.

INVENTORY SYSTEMS

Since inventory represents such a significant portion of the total assets of many firms, an inventory system should be an integral part of the accounting for inventories. Each system should be designed to meet the needs of the particular firm; however, all systems should be designed such that:

1. The information network is responsive to changing economic conditions.
2. Answers are provided for the basic questions of: how much to buy or produce, what items to purchase, and when to purchase the items?

For each firm, the inventory system will vary because of the:

1. Number of items in the inventory.
2. Management by owners or by appointed managers.
3. Importance of certain items in product line.
4. Value of inventory.

A firm with a limited number of items operated by the owners would not adopt an elaborate system to facilitate control. The controls are inherent because the owners are also the managers. As the size of the business increases, the owners may no longer be able to control directly the day-to-day operations; therefore, some additional controls must be employed.

The sophistication and degree of control should not outweigh the benefits to be derived. As an example, a firm should not use the Economic Order Quantity model (see Appendix to Chapter 7) for every item in the inventory. The costs involved in determining the inputs for the EOQ model would be prohibitive.

The effect of these considerations is evident in the choice between perpetual or periodic inventory methods. The perpetual method provides more information but it also costs more than the periodic method. The higher the value of the item, the more desirable the perpetual method. Consequently, the choice of the inventory system is a function of many variables, i.e., degree of control desired, value of inventory, cost of information, etc.

Types of Systems

Inventory systems run the gamut from signing delivery tickets and stocking the merchandise to a daily computer printout indicating the amount of inventory on hand. A corner grocer's inventory system would be much like the former, while a toy manufacturer's would typify the latter.

The corner grocer acknowledges receipt of the goods, stocks the items, and then records the sales price when they are sold. As the supply runs low, he calls his supplier and has another order delivered. Since the supplier probably has restrictions on the size of the order, the grocer must delay placing an order until reaching the minimum required. Thus, the grocer does not need EOQ models or computers to control his inventory. In his case employing sophisticated techniques would be a waste of money.

The toy manufacturer, on the other hand, may need inventory information on a daily basis. If the inventory is not moving in Atlanta but is selling in New York, information on a daily basis will enable management to take advantage of the market situation. This phenomenon is particularly crucial with a fad or fashion item. The goods must be in the right place at the right time. The widespread use of computers has given more firms the opportunity to maintain daily records of inventory.

In summary, inventory systems vary from firm to firm. Whether the information is recorded on delivery reports, computer punched cards, or merely entrusted to the memory of the owner is a function of many variables. Some of the pertinent questions that must be answered by any inventory system include:

1. When, from whom, and how much should be purchased? (Purchase order)
2. Was the merchandise delivered and recorded in stock? (Receiving report)
3. When was the merchandise sold and taken out of stock? (Sales invoice)
4. How was the merchandise shipped? (Shipping order)
5. Is the merchandise still in stock? (Inventory count)

SUMMARY

Inventories are an important part of the current assets of manufacturing and retailing firms. Consequently, management must exercise control not only for safeguarding the assets but also for insuring an adequate, though not excessive, supply of goods. The accounting function involves accumulating costs of the inventory and allocating these costs to cost of goods sold and inventory on hand. This allocation is made using one of several assumptions relative to cost flows, i.e., LIFO, FIFO, and weighted average.

APPENDIX TO CHAPTER 7

Planning and Control for Inventory

Planning and control are particularly important for inventory because:

1. Inventory represents a significant portion of the total assets of a firm.
2. Inventory is related to production.
3. Sales volume is limited by the quantity of inventory available.

The larger the inventory level, the greater the cost of maintaining the inventory. Increased levels of inventory require additional warehouse space, more labor hours, higher taxes, larger insurance coverage, etc.; however, the problem of inventory control is made easier because inventory is related to production. The amount of inventory can be controlled by regulating production. In addition, the costs involved in inventories are more identifiable than they are in some other areas, thereby leading to more quantifiable data. The more quantitative the information, the easier is its utilization in mathematical formulas to obtain more reliable and accurate answers. Consequently the inventory area is often the first area in the firm in which mathematics is employed.

Costs of Inventory

The major objective of inventory control is the minimization of total costs. There are two types of costs involved — ordering costs and carrying costs.

Ordering Costs

The costs of ordering are the expenditures involved in getting the item into the firm's inventory. They are incurred each time an order is placed and are expressed in terms of dollars per order. These costs must be determined from the labor, materials, etc. involved in issuing a purchase order, the receipt of the goods, and the processing of the payment to suppliers. Salaries will constitute the largest part of these costs.

Carrying Costs

The costs of carrying an item in inventory are those expenditures that must be made because the firm maintains an inventory. These costs include:

1. Interest on the money invested in inventory.
2. Warehouse costs such as heat, light, and power.
3. Record keeping for the inventory and the cost of taking the inventory.
4. Taxes and insurance on inventory.
5. Depreciation of facilities.
6. Obsolescence factor involved with the inventories.

Because of the difficulty of placing a dollar value on each of the particular costs included, carrying costs are generally presented as a percentage of the average inventory value. This value is chosen because, over time, a firm does not have the maximum or minimum amount of an item in the inventory but rather an average. The easiest method of determining average inventory is to add the opening inventory to the ending inventory and divide by two.

In addition to ordering and carrying costs, the firm must be aware of the loss in sales which may result from a failure to have an item in inventory (stockout). To minimize this cost, safety stocks are maintained for the more crucial items in the inventory. The more important the item, the larger the safety stock necessary. For example, a supermarket would keep a larger safety stock of coffee than of gourmet foods. For many customers, the lack of coffee in one store may encourage them to buy their groceries for the week in another store. Therefore, in this situation, coffee is a very important item in the product line of a supermarket.

Economic Order Quantity (EOQ)

The EOQ model is a mathematical formula used to attempt to minimize the total costs of inventory. Mathematically, the formula provides management with the quantity of the item that should be purchased. In addition, the reorder point can be ascertained by determining the usage of items. The use of a model does not, however, guarantee accuracy of the assumptions. If the cost information used in the formula is incorrect, the economic order quantity is also incorrect.

Economic Order Quantity Formula

$$EOQ = \sqrt{\frac{2RS}{C \times I}}$$

where

R = Total number of units required in a period

S = Cost per order expressed in dollars

C = Cost per unit

I = Carrying cost expressed as a percentage of average inventory

EXAMPLE. XYZ Company has decided to establish a control policy to reduce inventory costs. The firm estimates that a safety stock for Item A of 600 units is essential and that a three-week delivery schedule can be expected. The price per unit is \$4 and the processing cost per order is \$160. The annual requirements for Item A will be 26,000 and carrying costs will equal 13% of average inventory value. The economic order size is calculated as follows:

$$EOQ = \sqrt{\frac{2RS}{C \times I}} = \sqrt{\frac{2 \times 26,000 \times 160}{4 \times .13}} = \sqrt{\frac{8,320,000}{.52}} = \sqrt{16,000,000}$$
$$= 4000 \text{ units per order}$$

The next question to be asked is when the order should be placed? The reorder point is determined as follows:

$$\text{Reorder point} = \text{Average usage} \times \text{lead time} + \text{safety stock}$$
$$= \frac{26,000 \text{ units}}{52 \text{ weeks}} \times 3 \text{ weeks} + 600 \text{ units}$$
$$= 2100 \text{ units}$$

Thus, when the inventory level of Item A reaches 2,100 units, a purchase order for 4,000 units should be initiated.

The examples used for the EOQ model and reorder point are highly simplistic. The most difficult aspect in the use of the EOQ model is the determination of the costs that are used in the calculation. This is an area in which accounting information is crucial to management for appropriate decisions. If the data used as input are not relevant, then the answer generated by the model may lead to erroneous decisions.

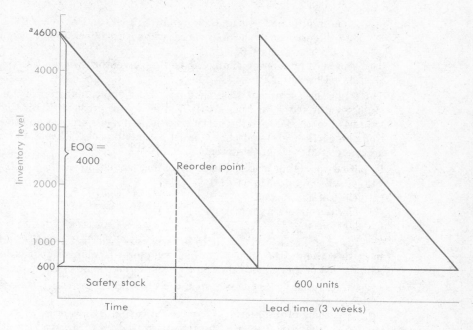

^aNote the initial quantity of inventory is 4600 units which is the EOQ of 4000 units plus the safety stock of 600 units.

QUESTIONS

7-1. Why is inventory a more important asset to a manufacturing or retailing firm than to a service organization?

7-2. What are the two major objectives concerning inventory, and what information is necessary to attain these objectives?

7-3. Distinguish between a periodic and a perpetual inventory system and give an example of a particular company which might be expected to use a periodic system and one which might be expected to use a perpetual system.

7-4. Company A, using a periodic inventory system, purchases $500 of merchandise from Company B. Credit terms are 2/10, n/30, and the goods are shipped F.O.B., Seller's plant. Freight charges amount to $50. What is the total cost of this purchase to Company A:
   (a) If they pay within the discount period?
   (b) If they do not pay within the discount period?

7-5. Explain the Discounts Lost account, indicating how it arises and the purpose served in terms of management information.

7-6. (a) If ending inventory is understated by $10,000 on December 31, 1974, and the error is never discovered, what is its effect on net income in 1974, 1975, and 1976?

(b) Would your answer change if ending inventory was initially overstated by $10,000?

7-7. The following statement was made by I. M. Wright, manager of the Sky High Food Mart: "Our business is limited to use of the FIFO inventory valuation method due to the perishable nature of our product and the fact that the earliest units purchased must be the first sold." Comment on the accuracy of Mr. Wright's statement.

7-8. In periods of rising prices, which inventory valuation method—LIFO or FIFO—will produce the largest:

(a) Cost of goods sold

(b) Ending inventory

(c) Net income

7-9. The gross profit method and the retail inventory method are two methods used to estimate the cost of ending inventory. Explain the situation in which each of these methods would be used and the procedures involved.

7-10. Inventory systems vary based upon the type of firm for which the system is designed.

(a) List some of the reasons for different inventory systems.

(b) Why would you expect the inventory system of a small firm to be different from that of a large firm?

7-11. Which type of inventory system is more costly to operate, perpetual or periodic? Why? Indicate a situation where the cost of maintaining a perpetual inventory method might be greater than the benefits derived from such a system.

7-12. What are the pertinent questions to which any inventory system must supply answers?

7-13. (a) Explain the terms *ordering* and *carrying* costs as they apply to inventory purchases (see Appendix to Chapter 7).

(b) What specific costs are included in each?

(c) A company requires 12,000 units per year and currently orders 1,000 units 12 times a year. What effect would a reduction to 6 orders of 2,000 units per order have on total ordering and carrying costs? Explain.

7-14. The economic order quantity (EOQ) model is an attempt at minimizing the total cost of inventory through a mathematical formulation (see Appendix to Chapter 7).

(a) What are the components of the EOQ model?

(b) What is meant by safety stock and why is it necessary to include the concept of safety stock in an EOQ model?

7-15. The following data apply to a particular inventory item of the Ken Company:

Safety stock = 500 units; two-week delivery schedule

Price per unit = $3

Cost per order = $150
Requirements per year = 25,000 units
Carrying costs = 10% of average inventory
(a) What is the economic order size?
(b) What is the reorder point? (Carry computations to nearest whole unit.)

PROBLEMS

Procedural Problems

7-1. The Lakeside Company sells two products, fins and shells. The sales data for the two products are presented for the month of May:

	Fins	Shells
Selling price per unit	$8.00	$12.00
Beginning inventory	200 @ $4.00	150 @ $8.00
Purchases:		
May 5	100 @ $4.50	200 @ $9.00
May 14	200 @ $5.00	100 @ $9.20
May 24	100 @ $5.50	100 @ $9.50
Ending inventory—units	150	200

Assume all expenses, other than cost of goods sold, amount to $2,000. Prepare two income statements for the month of May, one using LIFO inventory and the other using FIFO inventory. Present all supporting computations.

7-2. The beginning inventory and purchases made by the Mac Company during the month of January are shown as follows:

January 1 Balance   100 units @ $10
        10 Purchased 200 units @ $11
        17 Purchased 200 units @ $12
        22 Purchased 100 units @ $13
        28 Purchased 200 units @ $15

During the month of January, Mac Company sold 600 units of inventory.
REQUIRED
(a) Compute the amount of ending inventory for Mac Company using the LIFO, FIFO, and weighted average inventory methods.
(b) Which method would produce the largest profit for Mac Company?
(c) If Mac Company is interested in having its accounting records adhere to the matching concept, which inventory method would they most likely use?

7-3. On December 31, 1974, the Precious Metals Corporation counted their physical inventory and found $5,000 merchandise on hand. This was far below what management had expected, so further investigation was made.

Through the investigation, management discovered that the night watch-man had cooperated with thieves to loot the warehouse. Since the company carries insurance for such occurrences no loss was borne by the corporation. The night watchman was hired in January, 1974; therefore, assume that all thefts occurred during 1974.

REQUIRED. The insurance company has asked you to compute the amount of the loss attributable to this theft, based upon the following correct account balances compiled by the company accountant as of December 31, 1974:

Sales	$90,000
Purchases	50,000
Purchase returns and allowances	4,000
Freight cost (paid by Precious Metals)	3,000
Inventory 12/31/73	30,000

During the past ten years Precious Metals Corporation has had a consistent gross profit rate of 30% of sales, and there is no reason to believe a change will occur this year.

7-4. The Duds Department Store has two departments: Men's Clothing and Women's Clothing. Management of the store does not take a monthly physical inventory, but rather computes an estimate of inventory at the end of each month, using the retail inventory method. The departmental inventories are computed separately and then added together for income statement purposes. The following data relate to each department for the month of November:

	Men's clothing		Women's clothing	
	Cost	Retail	Cost	Retail
Beginning inventory	$ 6,000	$ 9,500	$ 5,000	$12,000
Purchases	28,600	44,500	22,900	46,500
Purchase returns	1,000	1,500	300	1,000
Sales		42,000		44,000

REQUIRED
(a) Compute the ending inventory by department at both cost and retail.
(b) What was the estimated gross profit for the store during the month of November?
(c) List some of the reasons a large department store would have for using a retail inventory method.

7-5. The Sole Brothers Shoe Company has determined the following costs pertaining to shoe leather inventory from accounting and production data. Cost per order is $40, and the cost of leather is $2 per pound. The Company uses 10,000 pounds of leather per year and their carrying costs amount to 10% of average inventory.

REQUIRED
(a) What is the economic order quantity?
(b) What is the optimum number of orders per year?

7-6. The beginning inventory of JMC Corporation consisted of 1,500 units costing $1,680, while their ending inventory totaled 3,800 units. The purchases during the year were as follows:

Date	Quantity	Total cost
March 15	2,000	$2,280
May 21	2,400	2,784
August 16	2,100	2,499
October 10	2,700	3,267
December 18	2,500	3,075

REQUIRED. Assuming a periodic inventory method is used, determine the cost of JMC Corporation's ending inventory using the following methods:
(a) LIFO
(b) FIFO
(c) Weighted average
(Carry computation to nearest whole cent.)

Conceptual Problems

7-7. The Ekten Manufacturing Company has a monthly usage for part #421J of 125 units. Inventory carrying costs are 25% of average inventory and ordering costs are $15 per order. Each part costs $2.00, and the company orders parts based upon their current economic ordering quantity. The supplier of part #421J charges $95 for freight on each shipment. Freight costs are included as part of the unit cost. Recently the supplier of part #421J offered Ekten Company a freight charge of $122 per shipment if they increased their present orders by 200 units. This would allow Ekten Company to decrease the number of orders placed during the year; however, the increase in order size will increase carrying costs to 40% of average inventory. Ordering cost will remain the same per order. Beginning inventory was 185 units; ending inventory amounted to 115 units. The change in ordering quantity from the new proposal should not materially affect the number of units in beginning and ending inventory.

REQUIRED. The president of Ekten Company has asked you to make a report to the board of directors outlining the most economical course of action for the company to follow. Your presentation should include computations indicating:
(a) The present economic order quantity.
(b) The number of orders presently placed per year.
(c) The number of orders to be placed if they accept the supplier's proposal.
(d) The effect of the new proposal on ordering and carrying costs.
(e) Whether or not the Ekten Company will benefit from acceptance of the proposal.

7-8. Redwood Lumber Company purchased 70% of its inventory on terms of 2/10, n/30. Total purchases for the year amount to $1,688,900. The company failed to take the cash discount on 40% of these purchases upon which the discount applied. What effect did the failure to exercise the discounts have on Redwood's profit for the year?

7-9. The Good-Buy Meat Market has recently completed its first month of operations. The income statement presented here reflects the profit earned by Good-Buy during the period:

Sales	$3,000.00
Cost of goods sold (2300 lbs.)	1,700.00
Gross profit	$1,300.00
Other expenses	1,100.00
Net income before taxes	$ 200.00
Income tax — 50%	100.00
Net income	$ 100.00

Three purchases were made by the market during the month. Each purchase was for 1,000 pounds of meat with the price per pound being $0.60, $0.80, and $1.00 for each of the three purchases, respectively. At a meeting between the owner and his accountant, the owner noted that the business was fortunate in that they sold a perishable product and were thus required to account for inventory using the FIFO valuation method. He pointed out that if LIFO had been used the market would have suffered a loss during its first month of operations. Do you agree with the owner's opinion? Comment. Would use of the LIFO valuation method have adversely affected the cash flow of the business?

7-10. The Hanks Company suffered a fire in its warehouse on November 15, 1974, and the company's entire inventory was destroyed. A physical inventory had not been taken since December 31, 1973, at which time the inventory had a cost of $38,000. The company records, which are kept in the main office, were not destroyed by the fire. They indicated that the cost of the merchandise purchased up to the time of the fire was $178,000, $18,000 of which had not been received from the supplier. Purchase returns amounted to $2,000 during 1974. Sales recorded for the period from January 1 to November 15 totaled $260,000; however, a credit sale to the Lydia Company recorded on November 14 was due for shipment on November 16. The amount of this sale was $15,000. Hanks Company has had a consistent gross profit rate of 30% of sales for the past five years. There is no reason to believe this will change for the first 11½ months of 1974.

REQUIRED. Calculate the amount of inventory destroyed by the fire.

7-11. The Springate Company began operations on January 1, 1974, at which time they hired R. U. Good as head of the accounts payable department. Mr. Good had little experience in this area, but he appeared to be a conscientious employee interested in saving the company money. When the company began operations, its cash position was rather weak, but the local bank had allowed it rather extensive borrowing privileges. These privileges included the right to make short term loans at 8% interest. During the first six months of operations, Springate Company purchased $480,000 of inventory, all subject to discount terms of 2/10, n/30. Mr. Good found that in most cases cash was not available to take advantage of the discount; therefore, payment was usually made 30 days after the purchase. Eighty percent of the purchases were made in this manner. Mr. Good felt that he saved the company money because, as he said: "It would be foolish to borrow money at 8% interest to take advantage of a 2% discount." Do you agree with Mr. Good? Present a schedule indicating the amount of discounts lost and also show how much it would have cost the company to borrow the money in order to take advantage of the discounts.

7-12. A review of the previous three years' ending merchandise inventory reveals the following errors:

12-31-72	$1,500	Understatement
12-31-73	$2,800	Overstatement
12-31-74	$2,400	Understatement

Each error is independent, and reported net income for the three years was as follows: 1972 – $8,900; 1973 – $10,500; 1974 – $12,400.

REQUIRED

(a) Compute the correct net income for the three years.
(b) Prepare the journal entry necessary to correct the accounts at December 31, 1974:
   (1) Assuming the books have not been closed at December 31, 1974.
   (2) Assuming the books have been closed at December 31, 1974.
(c) Assume the errors listed in the problem were not discovered or corrected. For how many years subsequent to 1974 would reported net income be affected? (Overstated or understated and by what amount).

8

INVESTMENTS

EDUCATIONAL OBJECTIVES

The material in this chapter is designed to achieve several educational objectives. These include:

1. An understanding of the criteria used for classifying marketable securities as current assets and long-term investments as a noncurrent asset.

2. A knowledge of the technical procedures involved in recording investments in stocks and bonds.

3. An understanding of the criteria used to distinguish stock dividends from stock splits and the different accounting treatment required.

4. An insight into the concept of present value and its usefulness in accounting measurements.

5. An appreciation for the necessity of properly controlling investments to minimize fraud and embezzlement.

The term *investments* refers to the assets of a firm, primarily in the form of stocks and bonds of other companies. The accounting and methods for control for investments closely follow those procedures used in accounting for cash. Since stocks and bonds are securities which can be converted into cash, internal controls are essential.

Investments can be divided into two classes:

1. Marketable securities. Investments made to employ temporarily idle cash with the intention of converting the securities into cash whenever necessary. Since these investments are expected to be of a relatively short duration, they are considered to be current assets and are so shown on the balance sheet.

2. Long-term investments. Investments made with the intention of holding the securities beyond the current accounting period. Since they are expected to be held beyond the current period, they are considered to be noncurrent.

Objectives in the Purchase of Investments

Management may have one of several objectives in mind when investments are purchased. The overriding goal is to maximize liquidity—profitability. The longer the holding period of the investment, the higher the rate of return demanded by the investor. In order to receive this higher return the investor must hold the security; however, holding the security reduces liquidity. If the firm is interested in liquidity, government obligations would probably be purchased. These are generally one- to three-month notes which achieve the flexibility desired by the firm.

The investment philosophy being discussed is not that of a financial institution. In those firms, the investment activity represents a major facet of their operation. Therefore, they would have the staff to undertake major analyses of potential investments. Other types of firms, which make only a limited number of investments, may have neither the time nor resources to research an investment thoroughly. As a result, their investments will probably be in the more well-known and established firms.

A cash budget should have been prepared; therefore, management should know how much, and for how long, idle cash will be available. Consequently, their investment decision will probably be along these lines:

1. Temporarily idle cash. These funds are available for a limited period of time and would be used to buy government or other short-term investments. Thus the firm receives some return on its idle funds, but at the same time it does not lose its flexibility.

2. Cash available over and above current or future requirements. The firm would make some analysis (cost-benefit) of the most productive investment. This analysis could lead to an investment in additional fixed assets, increased inventory levels, or long-term investments in stocks or bonds.

Investments in Stocks

A firm purchasing stock, normally, would consider the stock to be short-term or current; however, if the purchaser attempts to exercise some control or influence over the other firm, the stock is considered to be long-term. The firm cannot exercise control without owning the stock.

Control does not necessarily mean ownership of more than 50% of the stock. The ownership of 5% of the stock of American Telephone and Telegraph might provide effective control. There is such widespread ownership of AT&T stock that 5% represents a significant number of shares relative to any other block of stock. Therefore, influence is a better term than control to denote long-term investments. A firm may desire to influence its suppliers, attempt to change a dividend policy, or, perhaps, the prices charged by a retailer. In these situations, the investment in stock would be considered long-term.

On the other hand, the purchase of a small number of the shares of any company listed on the major stock exchanges would, generally, qualify as a short-term or current asset. The purchase of 500 shares of IBM would not be an attempt to influence any of the policies of that company; however, the purchase of 500 shares of a company whose total shares outstanding were 2,500 could be considered an attempt to exercise influence.

Investments in Bonds

The majority of bonds are considered to be long-term investments with the exception of government obligations, which generally are short-term. The majority of individuals purchasing corporate or municipal bonds appear to do so with the intention of holding them until maturity. The purchase of United States government bonds is generally considered to be a use of temporarily idle cash and, consequently, a current asset.

The maturity period of government obligations enables managers to invest funds for short periods of time; however, the interest rate paid on these obligations is low in relation to rates on corporate bonds. This is another indication that government obligations are temporary uses of idle funds.

Corporate bonds, generally, have higher interest rates than government obligations and typically have longer-term maturities, e.g., 20-year bonds. The major purchasers of these bonds are financial institutions and other investment-oriented entities, i.e., banks, insurance companies, pension trust funds, etc. They do not need the fixed assets that manufacturing firms require because, to a great extent, the objective of their business is to earn a profit through investments. In addition, by making longer-term investments, these entities will have access to funds in future periods when they will need them for pension payments, insurance claims, etc.

ACCOUNTING FOR MARKETABLE SECURITIES

Stocks

Purchase of Stock. Stocks are recorded at the market price plus any brokerage fees.

Example. ABC Company purchases 200 shares of Zuma Industries at $35 per share plus brokerage cost of $120.

Marketable Securities — Zuma Industries	7,120	
Cash		7,120

Note that the brokerage fee on the purchase is an addition to the purchase price. The rationale for this is that the brokerage fee is a cost necessary to obtain the investment. Consequently, these fees are recorded as part of the cost of the investment.

Sale of Stock. The sale of Zuma at $40 per share with a brokerage cost of $150 is as follows:

Cash ($8,000 — $150)	7,850	
Marketable Securities — Zuma Industries		7,120
Gain on the Sale of Stock		730

Note that the brokerage fee for selling the stock is a deduction from the gross sales price. In selling the stock through a broker, the commission is deducted by the broker before the seller receives any cash. Thus the commission is treated as a reduction of the sales price rather than shown as an expense.

Cash Dividend. The receipt of any cash dividend is recorded as a debit to Cash and a credit to Dividend Income. For example, the receipt of a $0.20 cash dividend per share on 500 shares of stock is recorded as:

Cash	100	
Dividend Income		100

Bonds

Purchase of Bonds. The accounting for the purchase of bonds differs from the accounting for a stock acquisition, in part because bonds have specified interest rates and payment dates.

EXAMPLE.   ABC Company purchases ten bonds of Levin Lighting Company, par $1,000 at 98 on June 30. Interest is paid on June 30 and December 31 at the rate of 7%. The entry to record the purchase is as follows:

Marketable Securities — Levin Bonds	9,800	
Cash		9,800

Note that the bonds are recorded at 98% of the total par value. The quoted market price of bonds is a percentage of par, in this case, 98%. Normally, par is either $100 or $1,000. Even if the par value were not $100 or $1,000, the bonds would still be quoted as a percentage of par. Consequently, to determine the price paid for the investment, multiply par value by the quoted price.

A question may be asked: Why does the purchaser buy bonds at a premium or a discount? A premium or discount arises because of the relationship between the market rate of interest and the stated rate of interest on the bond. The bonds pay a particular rate of interest which is stated on the bond by the issuing firm. There is also a market rate of interest established for similar types of bonds. Similar means basically the same risk factors, constraints on the issuing firm, and length of time the bonds will be outstanding.

Calculation of premium or discount. If the stated rate of interest is equal to the market rate, the bonds would be purchased at par; however, the probability of the *fixed* rate shown on the bond and the *variable* rate in the market being the same, at the time the bonds are sold, is low. Since the stated rate cannot be changed to coincide with the market rate, the problem is resolved by the investors adjusting the amount they are willing to pay for the bond. The face value remains the same but the bond price is such that the stated amount of interest will yield the market rate of interest.

In order to illustrate this situation, assume that a $1,000 bond has a stated rate of 6% and that the market rate has become 7%. The stated interest amounts to $60 per year; therefore, investors would be willing to pay an amount such that $60 would be equal to 7% of the purchase price of the bond. The determination of this price is:

$$\text{Purchase Price} = \frac{\text{Interest received}}{\text{Interest rate}}$$

$$X = \frac{\$60}{.07}$$

$$X = \$857$$

The illustration is a simple explanation of how the purchase price is determined. The assumption, in the illustration, is that the bond has a perpetual life. Therefore, the annual interest received can be divided by the rate. In practice, bonds have a fixed maturity date; therefore, there are a limited number of periodic interest payments. The calculation of the purchase price must then be based upon the receipt of interest during the life of the bond. This makes the calculation more complicated because, although the payments are the same, the value of a dollar to be received in ten years is not the same as a dollar received today.

Concept of present value. The concept of present value is used in many areas of accounting. Therefore, an understanding of the concept is useful to the user of financial statements. Present value can be explained by relating it to compound interest since they are two sides of the same coin.

The reason for the statement that a dollar today is worth more than a dollar received in ten years, or even next year, is that the dollar received today can be invested. If the dollar is invested then interest is earned and the dollar is worth more at the end of the year than at the beginning. In addition, the interest earned in the first year would earn interest in any succeeding year. This is known as *compound interest.*

As an example of compound interest, assume that $1,000 is invested at 8% for five years. The value of the investment at the end of each year is as follows:

Year	Principal at beginning of year	×	Interest rate	=	Interest for year	Principal at end of year
1	$1,000.00	×	8%	=	80.00	$1,080.00
2	1,080.00	×	8%	=	86.40	1,166.40
3	1,166.40	×	8%	=	93.31	1,259.71
4	1,259.71	×	8%	=	100.78	1,360.49
5	1,360.49	×	8%	=	108.84	1,469.33

From the example, it can be seen that $1,000 today invested at 8% compounded annually would have a future value at the end of five years of $1,469.33.

The concept of present value is the opposite side of the preceding example. That is, the present value of $1,469.33 to be received at the end of five years is $1,000 if the desired rate of return is 8%. Therefore, an investment opportunity must be evaluated in terms of the desired rate of return and the timing of the future incomes in order to determine the present value of the future incomes. This is the procedure used in determining the market price of bonds. The bondholders will receive periodic interest payments plus the face value at maturity. The present value of each interest payment and the face value received at maturity must be calculated in order to determine the present value of the bond.

The many possible desired rates of return together with a wide variety of time periods make the mathematical calculations time-consuming; however, present value tables have been developed which show the present value of $1 received at the end of each year for numerous periods of time and rates of interest. These tables are readily available for determining the present value of an investment opportunity.

Purchase of bonds between interest payment dates. If the bonds are purchased at any time other than the interest payment date, the purchaser must pay the seller for the interest earned from the last payment date. In the previous example, if the securities were purchased at 98 on September 30, 1974, the transaction would require a payment for interest from June 30. The entry to record the transaction would be:

Marketable Securities — Levin Bonds	9,800	
Interest Receivable	175	
Cash		9,975

The interest is calculated as follows:

$$\text{Par value} \times \text{Interest rate} \times \text{Time} = \text{Interest}$$
$$\$10,000 \ \times \quad 7/100 \quad \times \ 1/4 \ = \ \$175$$

Interest is always calculated on the par value of the bond, not the amount of cash paid. In the previous example, the cash outlay included interest for three months. This portion of the cash outlay will be returned to the buyer when the semiannual interest payment is received. The firm paying the interest on the bond is not concerned with the ownership during a particular time period; therefore, payment will be made to the owner of the bond on the date the interest is to be paid. The new owner collects the full interest on the interest payment date; thus he is reimbursed for the amount of the interest purchased from the previous owner.

The entry to record the collection of the interest on December 31, 1974 would be:

Cash	350	
Interest Receivable		175
Interest Income		175

The credit to Interest Receivable is necessary to eliminate the interest purchased from the previous owner. The credit to Interest Income is the amount earned by holding the bond for three months.

Sale of Bonds. In the sale of a bond the succeeding purchaser will buy the interest, for the number of days since the last interest payment, from the seller. Therefore, before the sale transaction actually takes place, an entry should be made to bring the interest up-to-date on the books of the seller.

Continuing with the previous example, assume that ABC Company sold the Levin Lighting Company bonds to Darden Industries at 97 plus interest on March 31, 1975. The entries to record this transaction are:
Entry to bring interest up-to-date:

Interest Receivable	175	
Interest Income		175

Entry for sale:

Cash [175 + .97(10,000)]	9,875	
Loss on Sale	100	
Interest Receivable		175
Marketable Securities — Levin Bonds		9,800

The interest must be brought up-to-date to determine the value of the Levin bonds to ABC Company on the day the bonds were sold. In the example, ABC received $9,875 for bonds with a book value on March 31, 1975 of $9,975.

Balance Sheet Presentation

Short-term investments are presented on the balance sheet in the current asset section. They are listed at cost or market, whichever is lower. Whichever one is used, the other is shown parenthetically.

An argument can be made for presenting these securities at market value even if the market value is higher than cost. The reason is that marketable securities are considered to be near cash, and the historical cost of the securities may be irrelevant if the securities are to be sold in the near future. In addition, the market value of securities is more readily determinable than is the market value for most other assets. However, as discussed in previous chapters, accounting is primarily concerned with reporting historical cost, and an upward valuation for one asset would not be consistent with present generally accepted accounting. Thus the present procedure is to show the market value parenthetically under the assumption that such presentation is sufficient information for the readers of financial statements.

ACCOUNTING FOR LONG-TERM INVESTMENTS

Long-term investments are made with the intention of holding them for an extended period of time. Consequently, the argument for presenting long-term investments at their market values on the balance sheet is not as persuasive as for marketable securities since they are not expected to be sold at the present market price.

Stocks

The initial accounting for long-term investments in stock is the same as for marketable securities. The initial cost includes the market price plus any brokerage fees. After the initial recording, the accounting treatment depends upon the extent of the investment in a particular company. There are two possible alternatives:

1. Cost method
2. Equity method

Cost method

This method is used in those cases where the investment in another company is so small relative to the total amount of outstanding stock that the investor is presumed to have little influence over the investee. According to the Accounting Principles Board of the American Institute of Certified Public Accountants, an investment of less than 20% of the voting stock of an investee would indicate a lack of influence. The use of the cost method may be limited since a basic feature of long-term investments is influence, and the Accounting Principles Board has stated that if the investor has significant influence the equity method must be used. If the cost method is justified, any dividends received are recorded as an increase in cash and an increase in dividend income.

Equity Method

The Accounting Principles Board issued Opinion Number 18: "*The Equity Method of Accounting for Investments in Common Stock.*" In this opinion, "the Board concluded that an investment (direct or indirect) of 20% or more of the voting stock of an investee should lead to a presumption that in the absence of evidence to the contrary an investor has the ability to exercise significant influence over an investee." If a significant influence exists, the proper accounting would be to use the equity method. Under this method, the investor records his applicable percentage of profits or losses of the investee as an increase or decrease in the investment. This is in contrast to the cost method where only dividends are recorded as income. Under the equity method, the receipt of dividends reduces the investment.

As an illustration of these methods, assume that Smith Company purchased 25% of Talbott Refrigeration for $300,000. The net income for Talbott was $100,000 for fiscal year 1975 and dividends of $5,000 were paid. According to generally accepted accounting principles, the equity method should be used because of the 25% ownership; however, for illustrative purposes both the cost and equity methods will be shown using the same information.

	Equity			Cost	
Purchase of 25% interest	Investment in long-term stock — Talbott	300,000		Same entry	
	Cash		300,000		
Recording of net income (100,000 × 25%)	Investment in long-term stock — Talbott	25,000		No entry	
	Investment income		25,000		
Receipt of dividend	Cash	1,250		Cash	1,250
	Investment in long-term stock — Talbott		1,250	Dividend income	1,250

Bonds

The accounting for a long-term investment in bonds is basically the same as for a short-term investment in bonds. There is interest attached to the long-term bond and the purchaser buys interest from the seller if the purchase was made between interest payment dates. Interest must be brought up-to-date to determine the value of the security at the time of sale.

When bonds are held to maturity, the bondholder will receive the maturity or par value of the bond irrespective of the original price paid. Therefore, even though the purchaser acquired the bond at a discount (below par value) or a premium (above par value), only the par value will be received at maturity. If a bond is purchased at 96 and the par or 100 is received at maturity, the discount of 4% will result in a gain for the purchaser and must be recognized at some point. The purchaser may prefer to wait until maturity to recognize this gain, or he may recognize it at the time of purchase. However, the matching concept states that revenues should be related to the expenses of producing the revenues. Thus, to properly follow the matching concept, the revenue from the discount should be recognized over the length of time that the bond is held. A simple method of recognizing the gain would be to allocate the discount equally over the period of time the purchaser holds the bond. In accounting, this allocation procedure for bonds is referred to as *straight-line amortization*.

A more complex method of allocating the premium or discount is referred to as *present value amortization*. The total discount or premium is prorated over the life of the bonds on the basis of a constant rate rather than a constant dollar amount. This method is more theoretically correct, but it requires bond yield tables or present value tables. Therefore, the illustrations in this text will follow the straight-line amortization method.

Note that any premium or discount on short-term investments in bonds is not amortized. The reason for this is that the holding period of the bonds should be nominal and, therefore, amortization is unnecessary.

To illustrate the accounting for a discount, assume the following information: A ten-year bond, par $1,000, 6% interest was purchased at 90 on January 1, 1975 as a long-term investment. The income to be received over the life of the bond would be:

Interest 10 years ($60 per year)	$ 600
Plus: Principal repayment	1,000
Total to be received for 10 years	$1,600
Less: Amount paid	900
Total income on bond	$ 700

The interest of $600 is recognized over the ten years as interest income. The $100 discount is not received until maturity, but it is earned because the bond is held for ten years. Therefore, the $100 should be amortized over ten years.

The entry to record the purchase of the bond is:

Investment in Long-Term Bonds	900	
Cash		900

The entry to record the straight-line amortization of the discount at the end of each year is:

Investment in Long-Term Bonds	10	
Interest Income		10

Therefore, over the lifetime of the bond, the investment in Long-Term Bonds account is increased on the books of the firm from $900 to $1,000.

If the bonds are sold before maturity, an entry must be made to record the interest earned since the last interest payment date. In addition, an entry to record the amortization of the discount or premium, since the last amortization entry, must be made.

For example, assume that a firm purchased a Jones Furniture 20-year, 8%, par $1,000 bond on January 1, 1974 at 140. Interest is payable on January 1 and July 1. The bond was sold on October 1, 1977 at 135 plus interest. Entries for the amortization are only made at year end.

The entries to record the interest earned and the amortization of the premium on October 1, 1977 are:
Interest earned for ¼ year (July 1–October 1):

Interest Receivable	20	
Interest Income		20

Amortization of premium for ¾ year (January 1–October 1):

Interest Income	15	
Investment in Long-Term Bonds		15

Recording these two entries establishes the book value of the bond on October 1, 1977. The book value is determined by:

Initial investment in bonds		$1,400
Less: Amortization for 1974 = 20		
Amortization for 1975 = 20		
Amortization for 1976 = 20		
Amortization for 1977 = 15		
	75	
Value of bond at 10/1/77		$1,325

The entry to record the sale is:

Cash ($1,350 + $20 interest income)	1,370	
Investment in Long-Term Bonds		1,325
Interest Receivable		20
Gain on Sale of Bonds		25

Note that in the case of a premium, the amortization reduces the interest income. The reason is that the holder of the bond receives the par value, not the purchase price, at maturity. For example, this bond was purchased at 140; therefore, the determination of total income on the bond would be:

Interest 20 years ($80 per year)	$1,600
Plus: Principal repayment	1,000
Total to be received for 20 years	$2,600
Less: Amount paid	1,400
Total income on bond	$1,200

Total income divided by 20 years equals $60 income per year. This amount can also be determined from the interest received per year of $80, less $1/20$ or $20 amortization of the premium.

Balance Sheet Presentation

Long-term investments in stocks are shown in the noncurrent section of the balance sheet at cost. Long-term investments in bonds are also shown in the noncurrent section of the balance sheet at cost plus or minus the amortization of the discount or premium. If the current market value is below cost, and the decrease is viewed as a permanent reduction in the asset value, the investment should be reduced to market. The loss would be shown as a deduction on the income statement.

Consolidated Financial Statements

If the extent of investment in the stock of another company is such that the investor has control (greater than 50%), consolidated financial statements should be prepared. When this situation exists the investing company is referred to as the *parent company* and the investee company is called a *subsidiary*. Consolidated financial statements are considered necessary for a fair presentation of the parent company since it has a controlling financial interest in the other company. In effect, the subsidiary is a branch or a division of the parent; therefore, consolidated statements are more meaningful than separate statements.

Since the subsidiaries are considered to be branches, or divisions, all of the companies involved are viewed as a single economic entity without regard to their legal distinction. Thus, consolidated financial statements are necessary in order to present the financial information about the single economic entity. The consolidated financial statements thus prepared must eliminate the transactions between the affiliated companies. If these transactions are not eliminated, the consolidated statements are misleading for statement users. Just as a company does not earn a profit selling between divisions, a parent and a subsidiary cannot profit on transactions between themselves.

Any intercompany transaction should be eliminated from the consolidated statements. Among the more common intercompany transactions and balances that should be eliminated are: receivables and payables, sales and purchases, investment in the subsidiary, and interest and dividends.

The portion of the subsidiary that is owned by investors, other than the parent, is called the *minority interest*. The minority interest is reflected on the balance sheet of the consolidated statements as a special type of owners' equity, i.e., minority interest in the assets of the consolidated economic entity. The minority interest does not reflect any ownership in the parent since their investment is only in the subsidiary. The subsidiary will provide financial statements to its stockholders.

SOME ADDITIONAL ASPECTS OF STOCK

Cash dividends, stock dividends, and stock splits are three events which can affect the investment in stock whether short-term or long-term.

Cash Dividends

Cash dividends are distributions of cash resulting from profitable operations of a firm, expressed as an amount per share. Some stocks are purchased primarily for their dividends, i.e., utility stocks, whereas others are purchased for potential capital appreciation. However, most stocks, over the long run, are purchased for both dividends and capital appreciation.

Stockholders have no legal right to dividends until they are declared by the board of directors. The board cannot unfairly withhold dividends from the stockholders; however, unfairness is difficult to prove. The board's failure to declare dividends may be the result of contingencies which are unknown to the stockholders. In addition, the firm must have cash available in order to pay a cash dividend. The firm may have a substantial balance in its Retained Earnings account and still not have any cash. Retained earnings represent profits retained in the business and these profits are invested in assets; therefore, they may not be available for distribution.

When dividends are declared, an entry is made to record the receivable. For example, Ramar Safari Company owns 100 shares of U.S. Steel Company. A cash dividend of $0.50 per share is declared on March 17, 1975 payable on April 15, 1975. The entry by Ramar to record the dividend is:

Dividends Receivable	50	
Dividend Income		50

The entry to record the receipt of the payment is:

Cash	50	
Dividends Receivable		50

In practice, if financial statements are not prepared between the date of declaration and the date of payment, many firms wait until the dividend is paid before making any entry. In this case, the entry would be:

Cash	50	
Dividend Income		50

Stock Dividends

In the event that the firm elects not to pay a cash dividend, a stock dividend may be declared. This is a distribution of stock in the firm to the present shareholder without any additional payment. The accounting treatment for the distribution of a company's stock as a dividend depends upon the size of the distribution. Most distributions are relatively small; however, the accounting treatment is the same for any distribution of less than 20–25% of the outstanding stock. In accounting terminology, a distribution of less than 20–25% of outstanding stock is called a *stock dividend*. Any larger distribution is, in effect, a *stock split*.

The objective of a stock dividend is to distribute something tangible (a stock certificate) to the owners without changing the market price of the security. Realistically, the price should fall, but the number of shares issued in a stock dividend is considered such a small percentage of the total number of shares that the market price probably will not be affected.

The individual investor can exercise the option of selling the shares for cash or holding them until some future time. If the investor sells the shares he will decrease his percentage ownership in the firm. Therefore, if influence or control is desired, the investor should not sell the shares.

There are no accounting entries necessary for a stock dividend on the books of the investor; however, the investor's cost per share is reduced because the acquisition cost is divided by the total of the old and new shares. For example, if Hallmark Boats receives a 10% stock dividend on its 250 shares of Lear Company, originally purchased at $50 per share, the new cost per share is $45.45:

	Share		Price	Total cost
Original purchase	200	shares @	50 =	$10,000
20% Stock dividend	20	shares	=	–0–
Total	220	shares @	(X) =	$10,000
Net cost per share		$\dfrac{\$10,000}{220}$	=	$45.45 per share

Stock Split

A stock split is a distribution of more than 20–25% of the outstanding stock of a firm. The number of shares issued is of such a magnitude that their effect on the market price is evident. The increased number of shares will be divided into the expected earnings, thereby reducing the earnings per share and as a result, the market price of the stock. However, the reduced market price is likely to increase investment activity in the stock which may be beneficial to the investor.

Some stocks have a particular range of market prices and, if the price exceeds that range, trading activity may be reduced. As a result of splits, stocks move back into their normal trading range and their market price may increase due to supply and demand factors. Buyers of particular stocks may not be willing to pay $120 per share; however, if the stock is split two-for-one, they may be willing to pay more than $60 a share. Consequently, stocks that are approaching their upper trading range are sometimes good buys for speculative purposes.

There is no accounting necessary for stock splits on the books of the investor because they are, in essence, stock dividends. The shares received in a split reduce the cost per share of the investment. A split differs from a stock dividend in that the market price of the stock changes on a split but not on a stock dividend. Distributions of more than 20–25% of the total number of shares outstanding are considered to be the equivalent of a split.

Generally, all stock splits are downward, i.e., the number of shares is increased and the price of the stock is reduced. However, a *reverse stock split* is used by some firms to move the price of the stock into a higher or more prestigious market. For example, if a stock were selling in the range of $2–$3, many people would not purchase these stocks because they consider them to be speculative. If the firm had a reverse one-for-five split, for every five shares the stockholder owned, he would receive one new share. With fewer shares outstanding, the earnings per share would increase fivefold and the price of the stock should increase to a $10–$15 range. This stock might, therefore, become more attractive to investors.

SYSTEMS AND CONTROL

For most firms, other than financial institutions, investment transactions do not occur as frequently as many other transactions. Therefore, a formal system may not be warranted except in the case of financial institutions. The primary concern of management is that controls are adequate to prevent misappropriation.

The investments normally should be maintained in a fireproof safe or safe-deposit box. The investment records should be maintained by someone other than the custodian of the securities. The records should include all possible identification, such as name of security, dates of purchase, par value, number of shares of stock, and maturity dates of bonds.

All investments should be held in the name of the firm. Since many marketable securities are bearer instruments, i.e., securities which can be converted to cash by whoever has possession, extreme caution must be used in safeguarding them. If the investments are held in a safe-deposit box, two individuals should be required to be present before the box can be opened. In this manner, one person cannot withdraw securities from the box without proper authorization and a witness.

Any investment decision should be authorized in the corporate minutes. Because of the possibility of fraud, spot checks should be made periodically to ascertain the existence of the securities. If the procedures just described are followed, the probability of fraud will be minimized.

SUMMARY

From an accounting point of view, investments may be divided into two classes: marketable securities and long-term investments. These two classes are distinguished primarily by management intent. If the intention is to invest temporarily idle cash, these investments are considered to be marketable securities and are current assets. If the intent is to influence the activities of the investee, the investment is long-term and is classified as a noncurrent asset. The two major types of securities for investment are stocks and bonds. Stocks are accounted for by the cost method except when the investment is 20% or greater of the outstanding voting stock, in which case the equity method is used. Bonds are recorded at cost and any premium or discount on long-term investments is amortized over the holding period of the bond. Many securities are bearer instruments; therefore, the internal controls over investments are similar to those required for cash.

QUESTIONS

8-1. Name and define the two classes into which investments may be divided.

8-2. (a) Does a firm purchasing stock normally consider the investment, short-term or long-term?

   (b) Under what conditions would a firm consider stock to be a long-term investment?

   (c) List some situations in which a purchaser of stock might wish to exercise some control or influence over the other firm.

8-3. (a) Are the majority of bonds considered to be short-term or long-term?

   (b) What is the exception and what factors cause it to be an exception?

   (c) Who are major purchasers of corporate bonds?

8-4. (a) What is the overriding goal of management when investments are purchased? Explain.
(b) What tool does management use to determine how much, and for how long, idle cash will be available?

8-5. The Ballary Company purchases stock at 98 and bonds at 98. Does this mean they paid the same price for the stock and the bonds? Explain.

8-6. Defend the argument that marketable securities should be presented on the balance sheet at market value rather than historical cost.

8-7. How are brokerage fees recorded in:
(a) The purchase of stock?
(b) The sale of stock?

8-8. Record the following transactions in a general journal:
(a) Webb Company purchases 100 shares of Allied Associates at 20 plus brokerage cost of $20.
(b) It receives a cash dividend of $2 per share.
(c) It sells 50 shares at 24 with a brokerage cost of $24.

8-9. Record the following transactions in a general journal:
(a) The Wise Company purchases 100 bonds of the Pettigrew Oil Corporation, par $100 at 99 on May 1, 1974. Interest is paid on March 1 and September 1 at the rate of 6%.
(b) It receives semiannual interest payment on September 1.

8-10. (a) Why would a purchaser buy bonds at a premium or discount?
(b) The MAP Corporation is issuing bonds with a par of $100 and an interest rate of 6%. Similar bonds have been sold, and a market rate of 7% exists. Would you be willing to buy a MAP Corporation bond (a) at a discount, (b) at face value, or (c) at a premium?

8-11. Mr. B. Lever, your boss, cannot understand why you credit Interest Income for the amortization of the discount which resulted from purchasing long-term bonds for less than par value. Explain the reason for the accounting utilizing in your explanation a comparison of the following two situations:
(a) Purchase a $100, 7%, ten-year bond at par.
(b) Purchase a $100, 7%, ten-year bond at 99.

8-12. Under certain conditions, the Accounting Principles Board of the AICPA recommends the equity method of accounting for investment in stocks. What is the equity method and under what conditions should it be used?

8-13. Why are consolidated financial statements prepared?

8-14. One man said: "I purchase stocks primarily for dividends. Therefore, I always invest in a company which has a substantial balance in its Retained Earnings account." Criticize this statement.

8-15. Compare and contrast stock dividends and stock splits by discussing the following:
(a) Percentage of outstanding stock.
(b) Entries to be made.
(c) Objective or purpose of the dividend or split.

PROBLEMS

Procedural Problems

8-1. Vic Noto purchased 400 of the 40,000 shares of Simms Corporation common stock at 34½ plus brokerage fees of $120 on January 1, 1974 as a temporary investment of idle cash. On May 10, Simms Corporation declared and paid a cash dividend of 80 cents per share. On July 1, Noto sold 250 shares at 30. The broker deducted his fee of $70 and remitted the balance to Noto. On October 10, Simms declared a cash dividend of 50 cents per share to be paid on November 10. Noto sold the remaining shares on December 15 at 38 receiving cash, net of brokerage fees of $45.

REQUIRED. Journalize the preceding transactions on the books of Vic Noto.

8-2. Assume the same information as in Problem 8-1 except that (*a*) instead of 40,000 shares, Simms Corporation only has 1,000 shares of common stock outstanding. Simms Corporation has $10,000 net income during 1974. (*b*) Vic does *not* sell any shares on July 1 or on December 15.

REQUIRED. Journalize the transactions on the books of Vic Noto.

8-3. The Ballard Bat Company has idle cash which could be invested in marketable securities. Billy Ballard, chief financial officer and nephew of the president, purchased 100 bonds of the Nellie Nice Spice Company, par $1,000 at 104 on June 1, 1974. The interest rate is 6%, payable on April 1 and October 1. On June 1, 1976, Ballard Bat Company needed some cash to continue operations. Instead of borrowing, the Nellie Nice Spice Company bonds were sold at 99.

REQUIRED. Journalize all the entries between June 1, 1974 and June 1, 1976 relating to the investment including year end adjustments.

8-4. Lincoln Corporation purchased bonds with a face value of $50,000 for 102.5 plus accrued interest. The bonds have a 6% interest rate payable on March 1 and September 1. The purchase was made on July 1, 1973 and the bonds mature on March 1, 1983. Lincoln Corporation considers this to be a long-term investment.

REQUIRED

(a) Prepare journal entries relative to the bond purchase from July 1, 1973 through December 31, 1973.

(b) Assume the same information except that the bonds were purchased at 95.5. Prepare journal entries for the same period.

8-5. Larry Dennis purchased 500 shares of Benton Book Case Corporation stock at 80 as a temporary investment on May 13, 1974. The broker's fees amounted to $100. On June 1, the Benton Book Case Corporation declared a cash dividend of $2 per share to be paid on July 1. Dennis records declarations of dividends as receivables. On August 15, Benton declared and distributed a 5% stock dividend. On October 10, Dennis sold 200 shares of Benton stock at 78. The broker remitted the balance after deducting his fees of $40. Benton Book Case Corporation split its stock issuing two shares for every one share outstanding on November 21.

REQUIRED
(a) Journalize the foregoing transactions.
(b) What is Dennis' cost per share at the end of the year?

8-6. (a) In each of the following situations determine the amount investors would be willing to pay for bonds. Assume that the bonds have a perpetual life:
   (1) $1,000 bond with a stated rate of 5%; the current market rate is 8%.
   (2) $5,000 bond with a stated rate of 6%; the current market rate is 4%.
   (3) $1,000 bond with a stated rate of 7%; the current market rate is 8%.
(b) Betts Corporation purchased 500 shares of Jennifer Company common stock. The stock was purchased for $32.25 per share plus a brokers fee of $75. Six months after the purchase Betts Corporation received a 20% stock dividend. Two months later a $0.20 per share cash dividend was received. The stock then split two-for-one. Subsequent to the split 300 shares of Jennifer Company stock were sold for $21.00 per share.
   (1) Compute the income Betts Corporation has earned on the stock from dividends and the gain on the sales.
   (2) Determine the number of shares Betts Corporation still holds and its net cost per share.

Conceptual Problems

8-7. The How-dee Company, a small manufacturing company, has a two-man office force. Frank Fysh is the manager and spends most of his time reviewing production reports and supervising the shop foremen. Bob Byrd is responsible for the bookkeeping functions and collecting incoming cash, making deposits, and writing checks. How-dee Company has been fairly successful; therefore, Bob has invested some of the extra cash in bonds. Frank admits that he knows nothing about the stock or bond market; therefore, Bob does all buying and selling of securities. After several transactions, Bob prepares a summary journal entry. He keeps no other record of the bonds. Since there are only a few bonds, he keeps them in a file folder in his desk which is never locked. For ease of selling, Bob has all the securities, except those which are bearer instruments, held in his name.

REQUIRED. Discuss any weaknesses in the internal control over investments and suggest steps to correct those weaknesses.

8-8. The president of the company is discussing the accounting treatment of investments. He states that, the way he understands it, "net income is affected by the decision to treat the investment in bonds as either long-term or short-term. After all, if an investment is treated as long-term, any discount is amortized and increases income each year." Therefore, he recommends that all bonds purchased at a discount be treated as long-term and all bonds purchased at a premium be treated as short-term. By so doing, he contends, the company will show a higher net income over the life of the bonds.

REQUIRED. Develop an answer to the president and, as a part of the answer, include a demonstration based upon a $1,000, 5%, five-year bond purchased at 97 and redeemed at maturity.

8-9. T. Jones and P. Johnson both have cash available to invest in either stocks or bonds. Johnson is interested in a steady flow of income and is not concerned about the size of the income so long as it is certain. Jones, on the other hand, is interested in getting as large a return as possible even if it means taking a chance of suffering a loss.

REQUIRED. By discussing the characteristics of bonds and stocks, indicate which Jones and Johnson will probably buy.

8-10. According to Accounting Principles Board Opinion Number 18, the equity method of accounting for investments should be used when the investment in another company exceeds 20% of the outstanding voting stock.

REQUIRED
(a) Explain the equity method.
(b) Why is the equity method considered necessary for fair reporting?

8-11. The Ashbrook Company has recently begun investing excess funds in securities of other corporations. Its initial investment was 1,000 shares of $10 par value stock of the Monteleone Corporation. The total price paid for the stock, including brokers fees, amounted to $38,500. The Company foresees no immediate need for the funds and thus plans to hold the stock as a long-term investment.

During the first year after purchase, the stock increased in value by $2 per share and Ashbrook Company received a $1 per share cash dividend. At the end of the year the accountant for Ashbrook Company made the following entry for these events:

Cash	1,000	
Investment in Monteleone Stock	1,000	
Dividend Income		2,000

In the second year after purchase Monteleone Corporation declared and distributed a 10% stock dividend. At the time of declaration, the stock was selling at $43.50 per share and the following entry was made on the books of Ashbrook Company:

Investment in Monteleone Stock	4,350	
Dividend Income		4,350

During the third year of holding the stock Monteleone Corporation declared a two-for-one split. The stock was selling for $46.50 per share immediately prior to the announcement of the split. To record this transaction the following entry was prepared:

| Investment in Monteleone Stock | 51,150 | |
| Income from Stock Split | | 51,150 |

At the beginning of the fourth year the president of Ashbrook Company decides to sell one-half of the 2,200 shares of Monteleone stock at the current market price of $23.25. The accountant for Ashbrook Company strongly urges the president not to sell the stock because his accounting records indicate that a sale at this time would result in a loss of approximately $22,000.

REQUIRED

(a) Describe how the accountant arrived at his estimate of the loss.

(b) If the accounting had been handled properly what entry would be made for the sale and what would be the remaining balance in the Investments account?

8-12. (a) The Paris Company purchased 6% bonds of the Camen Corporation as a long-term investment. The bonds pay interest semiannually on March 1 and September 1. Paris Company purchased the bonds on May 1, 1974. The bonds are dated March 1, 1974 and mature on March 1, 1984. On September 1, 1974, the accountant for Paris Company made the following journal entry:

Cash	1,200	
Accrued Interest Receivable		400
Investment in Long-Term Bonds		32
Interest Income		768

REQUIRED. Prepare the journal entry made on May 1, 1974 for the purchase of the bonds. Include all computations.

(b) The accountants for the Bordeaux Company recently made an entry for the purchase of bonds. The bonds had a maturity value of $10,000 and were purchased for $9,600. The investment account was debited for $10,000 and included in the entry was a credit for $400 to an account entitled Gain on the Purchase of Bonds. The accountant explained his entry in the following manner: "The bonds were purchased as a long-term investment and will be held to maturity. Thus we are assured of receiving $10,000 so the gain is a result of the purchase at an advantageous price and should be recognized at that time."

REQUIRED. Explain whether or not you agree with the accountant's statement. Justify your position.

9

FIXED ASSETS, INTANGIBLES, AND NATURAL RESOURCES

EDUCATIONAL OBJECTIVES

The material in this chapter is designed to achieve several educational objectives. These include:

1. A knowledge of the nature of fixed assets and the expenditures to be included in the cost of the asset.

2. An awareness that, in accounting, the purpose of depreciation is to allocate historical costs to the periods benefitted.

3. A knowledge of the alternative methods of depreciation and the effect of the choice of the methods on the income statement and balance sheet.

4. An understanding of the difference between tangible assets, intangible assets, and natural resources, and the accounting required for each type.

5. An understanding of the effect of inflation on the validity of financial statements, particularly the presentation of fixed assets on the balance sheet.

The term *Property, Plant, and Equipment*, commonly labeled Fixed Assets, refers to those assets of a firm acquired for utilization over more than one period in the profit-generating activities of the firm and not intended for resale. This definition differentiates fixed assets from inventory. In manufacturing firms, fixed assets are used in the production of inventory and the sale of the inventory generates revenue. Nonmanufacturing businesses also have fixed assets which are used in the operation of the firm. As a general practice, fixed assets are not intended to be sold, and they are sold only when the cost of maintaining the asset exceeds the revenue generated.

Productive fixed assets are particularly important because, without adequate productive facilities, inventory will not be produced or it will be produced at a cost that would price the product out of the market. Assets such as an abandoned plant or land held for future use are not productive assets and are not considered to be fixed assets. If they are ultimately used in the productive process, they would then be placed in the fixed assets category. In the interim, they are classified as long-term investments.

COST

At Acquisition

There are many expenditures made when an asset is purchased, e.g., gross cost less the sales discount, direct labor involved in installing the machine, cost of trial runs, etc. The accounting problem is to determine whether the expenditures were made for the benefit of the asset or merely coincidental with the purchase and thus, expenses of the period. These expenditures are difficult to categorize and, as a result, accountants must make decisions regarding the appropriate classifications. The rule or guiding principle to be used in these cases is:

> Any expenditure necessary to render the asset suitable for its intended use and which is expected to benefit future periods is a cost of the asset.

The objective in the purchase of any asset is the operation of that asset at its efficient level. Therefore, any expenditure made to achieve this goal is a cost of the asset. Costs such as transportation paid to have the asset delivered, installation cost, supervisory costs during installation, material and labor cost of trial runs, etc. are all costs of rendering the asset suitable for its intended use. In addition to the expenditures mentioned, there are many more that will not be discussed. The general rule provides adequate guidance for the accounting treatment of any other expenditure.

The general rule is based upon the rationale that all expenditures which benefit future periods should be included in the total cost of the asset. If these costs were not incurred then the asset would not be as useful to the firm. Therefore, these expenditures benefit future periods in the same manner as any asset benefits these periods.

When the acquisition price includes more than one asset, the cost must be allocated to the various assets, e.g., building, land, and equipment purchased for one price. The allocation of cost is on the basis of the appraised value of the assets. For example, building and land were purchased for $150,000. The appraised values were $55,000 for land and $110,000 for the building. The allocation of the $150,000 to the land and building is as follows:

	Appraised value	Portion of total	Acquisition cost
Land	$ 55,000	1/3 ($150,000)	$ 50,000
Building	110,000	2/3 ($150,000)	100,000
Total	$165,000		$150,000

Two primary reasons for the allocation are (a) to show the cost by type of asset for balance sheet purposes and (b) the cost of assets with a limited life must be allocated to the periods which will receive the benefit. Land does not have a limited life; therefore, the cost cannot be allocated.

After Acquisition

Expenditures are made at varying times throughout the lifetime of the asset as well as at the time of initial purchase. These costs are incurred for a variety of reasons, but they can be separated into two categories, i.e., expenses and capitalizable costs.

Expenses. When expenditures are made to retard the natural deterioration of the asset, they are recorded as expenses. Such expenditures are expenses because the effect of the expenditure is consumed in the current period. Expenditures of this type are necessary to preserve the future benefit potential of the asset, e.g., preventive maintenance; however, since the expenditure is made on a regular basis, it should not be capitalized as an asset.

Capitalizable Costs. Some expenditures are capitalizable because they benefit future periods. They either increase production, decrease operating costs, or increase the efficiency of the fixed asset. These costs can be divided into (a) costs incurred to improve the original qualities of the asset or (b) costs incurred to restore the asset to its original condition.

In situation (a), the expenditure is recorded as a debit to the asset account. The rationale for this treatment is that the expenditure has improved or increased the asset beyond the level of its original condition and the debit to the asset reflects the increase. In situation (b), the asset condition has not been increased beyond its original state. The expenditure has restored part or all of the asset to its original condition, i.e., overhaul of a motor. The expenditure is made to restore the motor to its original condition, but it does not make the asset better than its original condition. The expenditure does, however, benefit future periods and should be capitalized. Since the asset is no better than its original condition, the expenditure is recorded as a debit to the Accumulated Depreciation account rather than increasing the asset cost.

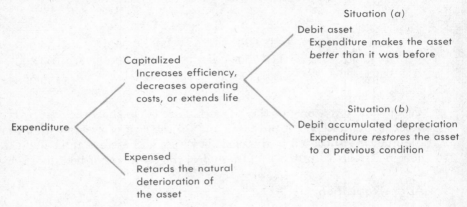

FIGURE 9-1

A comparison of the two methods of capitalization demonstrates that the effect on the amount to be depreciated is the same. Assume that an expenditure of $15,000 is to be capitalized, the two methods are:

	Before expenditure	Makes the asset better Situation (a)	Restores asset to original condition Situation (b)
Cost	$40,000	$55,000	$40,000
— Accumulated depreciation	20,000	20,000	5,000
Book value	$20,000	$35,000	$35,000

In both cases, the resulting book value (undepreciated cost) is $35,000. In situation (a), this was accomplished by increasing the asset cost from

$40,000 to $55,000 without changing the Accumulated Depreciation account. In situation (b), the accumulated depreciation was reduced from $20,000 to $5,000 while the asset cost remained the same. The entries for these two transactions are:

Situation (a)

Asset	15,000	
Cash or Accounts Payable		15,000

Situation (b)

Accumulated Depreciation	15,000	
Cash or Accounts Payable		15,000

DEPRECIATION

Nature

The cost of fixed assets is recorded on the books because future periods are benefitted. The allocation of the cost to future periods is called *depreciation*. Assets normally wear out or are consumed over varying periods of time. The reasons for their deterioration generally result either from consumption or obsolescence. The latter has become a significant factor since World War II because of technological advances. Firms are now more interested in increased capacity, speed, and size of fixed assets rather than mere physical existence or capability to produce.

Depreciation must be recorded because the usefulness of the asset to the firm is consumed over time. Therefore, a portion of the asset cost should be allocated to each period. Depreciation expense must be recorded to properly match the revenue earned through the use of the fixed assets with the consumption or usage of those assets, i.e., matching concept.

Each asset is a unique item and is affected by consumption and obsolescence in different ways. Therefore, it should not be surprising that there are different methods of depreciation. Some methods reflect the usage of the asset over time while others reflect projected obsolescence. The method selected should, to the extent possible, coincide with the usage or obsolescence of the asset. A method should not be selected because it maximizes any goal other than properly reflecting the usage of the asset. By selecting the appropriate method, both the objective of reporting for financial statements and the matching concept are satisfied.

Methods of Depreciation

Straight-line. This method assumes usage of the asset equally over time. The asset is considered to provide the same amount of benefits in each period. The calculations involved are:

1. Acquisition cost
   − Salvage value
   ——————————
   Depreciable base

2. $\dfrac{\text{Depreciable base}}{\text{Expected life}} = \text{Depreciation per time period}$

The acquisition cost is obtainable from the records of the company. The *salvage value* is the expected recoverable value of the asset at the end of its useful life to the firm. Thus this amount does not become an expense but reflects the return of a portion of the original investment; however, in the interest of clarity and materiality, if the salvage value represents a negligible amount it is frequently ignored. This practice does not satisfy the matching concept, but it does prevent the records from being cluttered by nominal amounts. In addition, the matching concept is not materially affected because the amount involved is insignificant.

Once the depreciable base is determined, it is allocated over the expected life of the asset. This estimate is normally made by an engineering approximation of the physical life of the asset. This is only an estimate which means that it could be wrong; however, it is assumed to be the best approximation of the expected life of the asset.

Example: XYZ Company purchased an asset for $13,000 cash on July 1, 1974. The expected life was five years and the salvage value $1,000. The depreciation for 1974 and 1975 is:

Acquisition cost	$13,000
− Salvage value	1,000
Depreciable base	$12,000

$$\frac{\text{Depreciable base}}{\text{Expected life}} = \frac{\$12,000}{5 \text{ years}} = \$2,400 \text{ per year}$$

Since XYZ did not purchase the asset until July 1, the depreciation for 1974 must reflect this fact. Therefore, the depreciation for 1974 is $1,200 ($2,400 × ½). The expense for 1975 would be for the entire year or $2,400. The entries would be:

12/31/74	Depreciation Expense	1,200	
	Accumulated Depreciation		1,200

and

12/31/75   Depreciation Expense   2,400
            Accumulated Depreciation   2,400

If at some future time the original estimate is determined to be incorrect, the new estimated life is used to calculate a new rate. This new rate is applied only to the remaining undepreciated amount and thus has no effect on prior periods. In accounting, revision of estimates is not considered an error in past reporting but rather effects only future statements.

Units of production method. This method assumes that the life of the asset is a function of the units it can produce. A machine has the capacity to produce a limited number of items; therefore, as an item is produced the total number of possible units is reduced. Consequently, depreciation may be charged on a per unit basis rather than on a time basis. Whereas straight-line depreciation is a fixed charge per time period, units of production depreciation is a fixed charge per unit of output.

EXAMPLE. XYZ Company purchased an asset for $13,000 cash on July 1, 1974. Assume that this machine could produce 60,000 widgets and at the end of its productive life would have a salvage value of $1,000. In 1974, 7,000 widgets were produced and in 1975, 18,000 were produced. The depreciation for 1974 and 1975 is calculated as:

Acquisition cost	$13,000
— Salvage value	1,000
Depreciable base	$12,000

$$\frac{\text{Depreciable base}}{\text{Units of output}} \quad \frac{\$12,000}{60,000} = \$0.20 \text{ per unit}$$

Depreciation for 1974 would be $0.20(7,000) = $1,400. The depreciation for 1975 would be $0.20(18,000) = $3,600. The entries are:

12/31/74   Depreciation Expense   1,400
            Accumulated Depreciation   1,400

and

12/31/75   Depreciation Expense   3,600
            Accumulated Depreciation   3,600

Sum-of-years'-digits. This is an accelerated method of depreciation which takes into consideration not only obsolescence, but also the increased usage of an asset during its early life. At the beginning of an asset's life, consumption is probably greater than at any other time.

This is attributable to fewer machine breakdowns and lower maintenance requirements. As the asset becomes older, the probability of down-time is greater and maintenance costs higher which results in the asset being less productive in its later life. Thus these factors provide the basis for higher depreciation charges when the asset is new and lower charges in later years.

The depreciable base in the sum-of-years'-digits depreciation is the acquisition cost minus salvage value. The term *sum-of-years'-digits* refers to the summation of the years' digits, which are used as the denominator for the allocation of cost. The numerators are the years in reverse order. Example, if an asset had a life of four years, the calculation would be:

$$
\begin{array}{l}
4 \\
3 \\
2 \\
\underline{1} \\
10
\end{array}
$$

10 = Total sum-of-years'-digits

1st year depreciation would be 4/10(X)
2nd year depreciation would be 3/10(X)
3rd year depreciation would be 2/10(X)
4th year depreciation would be 1/10(X)

10/10(X)

where X = Depreciable base

EXAMPLE. XYZ Company purchased an asset for $13,000 cash on July 1, 1974. The expected life was five years with a salvage value of $1,000. The depreciation for 1974 and 1975 is calculated by:

Acquisition cost	$13,000
— Salvage value	1,000
Depreciable base	$12,000

Sum-of-years'-digits

5	1st year depreciation 5/15 (12,000)=	$ 4,000
4	2nd year depreciation 4/15 (12,000)=	3,200
3	3rd year depreciation 3/15 (12,000)=	2,400
2	4th year depreciation 2/15 (12,000)=	1,600
1	5th year depreciation 1/15 (12,000)=	800
15		$12,000

The depreciation for 1974 is $4,000 × ½ = $2,000 and the entry is:

Depreciation Expense	2,000	
Accumulated Depreciation		2,000

The depreciation calculation for 1975 is more complex. In this method, the $5/15$th rate is for the period 7/1/74 to 7/1/75. Therefore, it is not permissible to use the $4/15$ rate for all of 1975. The depreciation for 1975 is $[(1/2)(5/15)(12,000)] + [(1/2)(4/15)(12,000)]$ or $\$2,000 + \$1,600 = \$3,600$. The use of two rates will always be required when an asset is purchased at any time other than at the beginning of the year and the sum-of-years'-digits method is used.

The sum-of-years'-digits can be calculated using the following formula:

$$\frac{N(N+1)}{2} = \text{Sum-of-years'-digits}$$

where N = Number of years

Using the previous example of five years:

$$\frac{5(5+1)}{2} = \frac{5(6)}{2} = 15$$

This formula is especially useful if the number of years involved is large, e.g., 25 years; however, none of the problems in this text will contain numbers large enough to warrant memorizing the formula.

Declining Balance. This method also takes into consideration increased usage in the early years of the asset's existence. In addition, since any accelerated method of depreciation increases depreciation in the earlier years, the effect of the asset becoming obsolete earlier than expected is minimized. The increased depreciation charge reduces the amount to be depreciated, thereby reducing the potential loss that may be sustained in a subsequent sale or trade-in.

The most common form of declining balance depreciation is *double-declining balance*, which means doubling the *straight-line rate*. The rate, not the length of time, is doubled. For example, if an asset has a ten year life, the straight-line rate would be 10%. The double-declining balance method doubles the rate (20%), and to calculate the depreciation the undepreciated cost is multiplied by the rate.

The cost is not reduced in this method by the salvage value as it is in the other methods. The application of a percentage to a declining balance provides for a residual amount. For example, if you walk one-half the distance to the nearest wall and then walk half the remaining distance, etc., you will never get to the wall. This example demonstrates the mathematics of the method and the source of the residual value.

EXAMPLE. XYZ Company purchased an asset for $13,000 cash on July 1, 1974. The expected life was five years and the salvage value $1,000. The depreciation schedule for the life of the asset is calculated as:

Straight-line rate = $\frac{1}{5}$ or 20%

Double this rate = 40%

1st year depreciation .40(13,000) = $5,200
2nd year depreciation .40(13,000 − 5,200) = $3,120
3rd year depreciation .40(13,000 − 5,200 − 3,120) = $1,872
4th year depreciation .40(13,000 − 5,200 − 3,120 − 1,872)
   = $1,123 (rounded)
5th year depreciation .40(13,000 − 5,200 − 3,120 − 1,872
 − 1,123) = $674 (rounded)

This method provides for a residual value of $1,011. As a practical matter, the depreciation in the last year is normally the amount necessary to reduce the undepreciated amount to the expected salvage value. Therefore, the amount of depreciation charged in the fifth year would be $674 + $11 or $685, and the undepreciated balance is the salvage value, or $1,000.

Evaluation of Depreciation Methods

A comparison reveals the differences between depreciation expense under each method. This comparison should point out that the choice of the depreciation method may have a significant impact on the amount of net income.

Year	Straight-line	Sum-of-years'-digits	Double-declining balance
1st	$ 2,400	$ 4,000	$ 5,200
2nd	2,400	3,200	3,120
3rd	2,400	2,400	1,872
4th	2,400	1,600	1,123
5th	2,400	800	685
	$12,000	$12,000	$12,000

The units of production method is not included in the comparison because it is a function of the units produced; these vary with each period of time. Note that the accelerated methods of depreciation (sum-of-years'-digits and double-declining balance) provide significantly more expense and therefore lower net income and lower taxes than does the straight-line method in years 1 and 2. Therefore, management may believe its best interest is served by choosing a particular depreciation method for reasons other than the matching of revenues with expenses.

Graphically, a comparison of the depreciation methods is as follows:

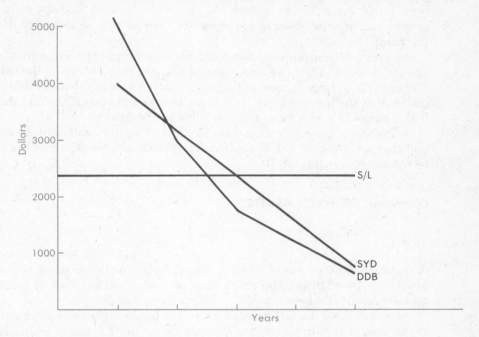

S/L—Straight-line depreciation
SYD—Sum-of-years'-digits depreciation
DDB—Double-declining balance depreciation

FIGURE 9-2

Note that under all three methods the total depreciation is $12,000. The difference is the allocation of the cost on a year-to-year basis. There are other methods of depreciation but the objective of each is the same: *to properly allocate the acquisition cost over the expected useful life of the asset.*

The choice of depreciation methods has an impact on the balance sheet in that the amount of depreciation expense increases the accumulated depreciation. Thus the net book value of the fixed asset is directly affected by the depreciation method; however, since book value is not intended to be current value, the effect of the depreciation method may not be significant in terms of the balance sheet. A point to be remembered is that, for accounting purposes, depreciation is a process of cost allocation, not valuation.

Another important aspect of depreciation is that the method chosen for book purposes does not have to be used for tax purposes. Thus an accelerated method may be used for tax purposes to reduce the current payment of income taxes while straight-line depreciation is used for book purposes. If the firm does use an accelerated method the reduced income tax has the effect of reducing the cash outflow in the early life of the asset.

As previously discussed, the total depreciation is the same over the life of the asset. Thus the total income tax and cash outflow is the same. The benefit of the reduction of tax in the early years is the availability of the cash at the present time. This is another illustration of the fact that a dollar today is worth more than a dollar in the future.

Since the different depreciation methods result in different incomes and taxes, a deferred tax is created. The accounting for deferred tax will be discussed in Chapter 10.

DISPOSAL OF FIXED ASSETS

Sale

A fixed asset may be sold when a newer, faster, or better asset becomes available. In addition, the asset may be sold when the cost of maintenance exceeds the benefits derived from its continued use.

An example of the latter situation might be one's personal automobile. If the cost of repairing the car exceeds the owner's finances or interest, he would probably attempt to sell the car. The sale of fixed assets is not unique to individuals; such transactions are common occurrences in industry. In selling the car the owner eliminates, totally, his equity in the car and receives the proceeds. In industry, entries must be made so that management and, ultimately, the readers of the financial statements, know what happened.

The guiding principle in the sale of a fixed asset is that all amounts associated with the item being sold are removed from the books. Similar to the situation of the individual, the firm eliminates its equity in the asset. Consequently, the asset and its related accumulated depreciation must be removed from the books.

To properly show the expense for the period and the related accumulated depreciation, an entry must be made prior to recording the sale, to bring the depreciation up-to-date. For example, if an asset was sold on July 1, 1975 and no depreciation had been recorded since December 31, 1974, it would be necessary to record one-half year's depreciation expense for 1975. The reason for the entry is a proper matching of the operating expenses and revenues of the current year. If the entry were not made, any loss incurred on the sale would be increased by the failure to record the expense and any gain would be reduced.

EXAMPLE. Dawson Company sold a machine for $5,000 cash. The original cost of the machine was $28,000 and it was 75% depreciated after the depreciation was brought up-to-date. The entry for the sale is:

Cash	5,000	
Accumulated Depreciation	21,000	
Loss on Sale of Machine	2,000	
Machine		28,000

Dawson sold for $5,000 a machine with a book value of $7,000 ($28,000 − $21,000). Consequently, a loss of $2,000 was incurred on the transaction. On the other hand, if the machine had been sold for $11,000, a gain of $4,000 would have been earned. The entry for this transaction is as follows:

Cash	11,000	
Accumulated Depreciation	21,000	
Machine		28,000
Gain on the Sale of Machine		4,000

The gain or loss on the sale of a fixed asset is placed on the income statement under the caption *Other Income and Expenses.*

Trade-In

In many cases, fixed assets are not sold but are traded in on newer models. The firm expects to continue operations and needs fixed assets to produce the product or products that it is in business to sell. Unless the firm is to be liquidated, or its scope of operation changed, the same type of assets are required.

There are three methods in accounting for trade-ins. Under each method, the asset to be traded in is removed from the books together with its related accumulated depreciation. In addition, a certain amount of cash or its equivalent, i.e., notes payable, accounts payable, etc., is given up in the exchange and must be credited. The three methods are: list price, tax, and secondhand market value.

List Price. This method records the new asset at its list price. The assumption is that the list price is a true reflection of the asset's worth. For example, the price for most industrial equipment is reflective of its worth and the list price method is acceptable. On the other hand, the list price method would not be acceptable for many new cars. The list price in this case is the starting point for negotiation.

EXAMPLE. Engle Company traded in Machine A on Machine B which had a list price of $30,000. Machine A originally cost $20,000 and was 60% depreciated after the depreciation was brought up-to-date. In addition to Machine A, Engle paid $25,000. The entry for the trade-in is:

Machine B	$30,000	
Accumulated Depreciation —		
Machine A	12,000	
Loss on Trade-In of		
Machine	3,000	
Machine A		20,000
Cash		25,000

The loss on the trade-in of $3,000 occurred because cash in the amount of $25,000 and Machine A, whose book value was $8,000, were given up in exchange for an asset with a list price of $30,000. The reason for the loss was that the book value of Machine A was not the same as its present market value.

Tax Method. The Internal Revenue Service requires the use of the tax method in accounting for trade-ins. Using this method, there are no gains or losses on the trade-in. Assuming the same facts as earlier, the loss would be treated as an increase in the cost basis to be used for tax purposes. Therefore, the loss will increase the amount of periodic depreciation to be charged in the future. A gain is the opposite: the cost basis is decreased, resulting in lower depreciation and increased net income. Thus the tax method requires that the loss or gain be spread over the life of the new asset rather than being taken in one year.

Utilizing the previous example, the entry under the tax method would be:

Machine B	33,000	
Accumulated Depreciation —		
Machine A	12,000	
Machine A		20,000
Cash		25,000

Note that the cost basis of Machine B equals the cash paid plus the book value ($8,000) of the asset traded. This is always the case using the tax method.

Secondhand Market Value Method. Both theoretically and practically, the secondhand market value is a more important aspect of a trade-in than the book value. The market value of the asset plus the cash paid constitutes the value of the total trade-in from the point of view of the firm. Thus, the secondhand market value method requires recording the new asset at the cash paid plus the secondhand market value of the asset traded.

The difficulty with the secondhand market method is that many assets are not readily salable and no market value can be determined. If, however, a secondhand market value can be obtained, and is reflective of the assets' value, this method should be used for book purposes.

From the previous example, if the secondhand market value of Machine A was $6,000, the entry for the trade-in would be:

Machine B	31,000	
Accumulated Depreciation—		
Machine A	12,000	
Loss on Trade-In of Machine	2,000	
Machine A		20,000
Cash		25,000

The loss on the trade-in is the difference between the book value of $8,000 and the market value of $6,000. Note that the cost basis of Machine B is the cash paid ($25,000) plus the secondhand market value of the old asset ($6,000).

Disclosure

The discussion concerning fixed assets has proceeded from the acquisition cost, costs incurred after acquisition, the allocation of cost to the periods benefitted, or depreciation, to the subsequent sale or trade-in of the asset. From the viewpoint of the readers of financial statements, disclosure of this information is necessary in making investment decisions. The questions of how and what to disclose must be answered.

First, the expense is listed in the income statement either as a cost of products produced, i.e., Cost of Goods Sold section, or as an operating expense. The asset and the related accumulated depreciation are shown on the balance sheet as:

Fixed Assets		
Machine A	$25,000	
Less: Accumulated depreciation	10,000	
Book value		$15,000

In addition to the cost and related depreciation, the depreciation methods employed should be disclosed. This enables the reader to note the different methods of depreciation used and the effects that these methods will have on future net income. This information is required by

Accounting Principles Board Opinion Number 22 which is more fully explained in the section on the disclosure of intangibles.

Effects of Inflation

Accounting data are measured on the assumption of a stable dollar; however, in the economy of the United States and most of the world, this assumption is not entirely true. In the United States, inflation has occurred on a rather continuous basis since 1940. Inflation means that the purchasing power of the dollar has declined. Thus, as commonly stated, if inflation is 5%, it would take $1.05 to purchase the same quantity of goods and services which could have been purchased previously for $1.00.

The changes in the purchasing power of a dollar are now ignored in conventional financial statements. The reason usually given for ignoring inflation has been that the extent of inflation has not been significant enough to warrant adjusting these statements. This means that the cost of assets purchased in 1945 is added to the cost of assets purchased in 1965 on the financial statements in 1974. As an example, an asset which cost $600 in 1945 might have cost $1,100 in 1965. Therefore, if a firm purchased an asset in each of these two years, the amount shown on the balance sheet would have been $1,700.

The problems of inflation are particularly significant in the accounting for fixed assets because of the long life of these assets. The current assets are constantly changing so that the impact of inflation on the balance sheet presentation of current assets is not as severe as with long-term assets.

An index of the change in purchasing power may be used to adjust historical cost financial statements for the effect of inflation. Many accountants and the AICPA have recommended that price level adjusted statements be presented as supplements to the usual financial statements.

Few companies have, however, provided this supplemental information. The reasons frequently cited for the failure to provide this information include:

1. There is no single price level index applicable to all assets.
2. The readers of the financial statement might be misled.
3. The rate of inflation has not been great enough to warrant adjustments.

INTANGIBLES

Intangible assets are defined as economic rights and advantages which will generate revenue in the future but do not have the physical characteristics of tangible assets, e.g., patents, trademarks, copyrights. Because tangible assets are visible it is easy to know when they exist; however, evidence of intangible assets is often vague and difficult to determine. In addition, the time lag between the expenditure and the development of an intangible asset is such that the expenditure might be expensed at its incurrence.

The accounting for intangibles is the same as that for tangible assets. Any expenditure necessary to render the asset suitable for its intended use and which is expected to benefit future periods is a cost of the asset. If a firm purchases an intangible, the cost to be recorded is the purchase price. The accounting problem occurs when a firm develops an intangible asset. This development may extend over many years and involve several people. The accounting problem is one of what to do with the developmental costs as they are incurred. If the accountant and management are reasonably certain that a valuable process or idea which can be patented or copyrighted will be developed, the costs should be capitalized, but given the nature of most research, it would be rare for the accountant to know that an asset would evolve.

Any intangible asset that is developed must be technically as well as economically feasible. There is little value in developing a product that cannot be sold in a competitive market. Therefore, unless the item being developed can be produced and sold, it cannot be considered an asset and no costs should be capitalized.

From the preceding discussion, it can be seen that developmental costs present substantial problems for the accountant. In addition, management is concerned with minimizing taxes and the write-off of expenditures in the period incurred reduces net income and taxes payable. As a result of the problems in properly accounting for research and development expenditures, the Financial Accounting Standards Board issued its Standard Number 2 which states that all research and development expenditures should be written off in the period incurred.

Amortization of Intangibles

The cost of intangibles must be allocated over the periods benefitted when the developmental costs can be associated with an asset or when intangibles are purchased. This process of allocation is called *amortization*.

There are two types of intangibles:

1. Some assets have a life limited by law or the nature of their existence, e.g., patents (legal life 17 years), copyrights (legal life 28 years, renewable for 28 more), leases, licenses, and goodwill of a limited duration. *These assets should be amortized over the shorter of their legal life or economic life.*

2. Some assets have an indeterminate life, e.g., trade names, secret processes, organization costs, and certain types of goodwill. Theoretically, if these assets are not consumed they should not be amortized. However, the Accounting Principles Board in its Opinion Number 17 stated that all intangibles must be amortized over a period not to exceed 40 years. Therefore, all intangible assets must now be amortized irrespective of the fact that they may have unlimited lives.

Theoretically, the amortization of intangibles should be systematic and reflect, over time, the decline in economic potential. According to the Accounting Principles Board Opinion Number 17, "the straight line method of amortization should be applied unless a company demonstrates that another systematic method is more appropriate." For example, if a patent, which cost $5,000, was expected to be productive for five years, the amortization entry for the year on a straight-line basis would be:

Patent Expense	1,000	
Patent		1,000

Unlike tangible assets, the credit is normally to the asset account rather than a *contra* account. Also, the amortization expense may be either a manufacturing or selling and administrative expense, depending on the nature of the intangible.

Disclosure

The Accounting Principles Board in its Opinion Number 22 stated that information concerning the accounting policies adopted by a reporting entity is essential for financial statement users. Included among the policies to be disclosed are those relating to the amortization of intangibles. In general, the disclosure should encompass important judgments as to the appropriateness of principles relating to recognition of revenue and allocation of asset costs to current and future periods. The Board suggests that adequate disclosure could be given in a separate *Summary of Significant Accounting Policies* preceding the notes to financial statements or in the initial footnote.

NATURAL RESOURCES

Natural resources are those wasting assets that are reduced as a result of the removal or sale of these resources. Examples of such assets are: minerals, timber, oil, gas, etc. Natural resources differ from depreciable assets in that they are consumed, i.e., their physical appearance is changed and diminished. Depreciable assets, usually, do not change their physical appearance as they are used. The accounting problem is the allocation of the cost of the resource to the periods benefitted. This allocation process is called *depletion*.

As with depreciable assets. the cost basis for depletion must be determined. The acquisition cost includes the purchase price plus any developmental or exploratory costs incurred. This amount is then reduced by the residual value of the property after the natural resource is removed. After the minerals, timber, etc. have been removed, the expected value of the land constitutes the residual value.

EXAMPLE

Acquisition cost	$2,000,000
Less: Residual value of property	200,000
Depletable base	$1,800,000

The method of depletion is, generally, on the units of production basis. As each unit is removed the value of the resource is reduced by that amount. The only difficulty with this method is forecasting the recoverable units. Normally this can be accomplished through engineering or geological estimates. In the previous example, if the estimated recoverable units were 900,000, the depletion per unit of output would be computed as follows:

$$\text{Cost per unit} = \frac{\text{Depletable base}}{\text{Total recoverable units}} = \frac{\$1,800,000}{900,000} = \$2 \text{ per unit}$$

The cost per unit is then multiplied by the number of units extracted during the period. For example, if 70,000 units were extracted in 1974, the entry for depletion would be:

Depletion Expense	140,000	
Natural Resource		140,000

As in the case of intangibles, the credit is normally to the asset rather than the *contra* account.

Disclosure

The undepleted cost of the natural resource should be shown on the balance sheet. Disclosure in a footnote of the total number of estimated recoverable units is appropriate so that the readers have an approximation of the amount that has already been recovered. The depletion cost is normally a production cost, and it is included as a part of the Cost of Goods Sold section of the income statement.

FIXED ASSET SYSTEM

This discussion will cover only a system for fixed assets. For most firms, intangibles and natural resources will not be of sufficient magnitude to warrant the development of a special system of controls. On the other hand, most manufacturing firms will have sufficient fixed assets so that control is vitally important.

Investment Decisions

A system of controls begins with the initial consideration to purchase a fixed asset. A capital budget, or forecast of fixed asset needs, should be prepared to determine the priorities for the use of the firm's scarce resources. The accountant should aid management in the preparation of the capital budget and also in the development of financial data needed to make decisions from among the various alternatives.

Each alternative must be viewed in light of its expected return to the firm. Management should not make an investment unless the asset will contribute something to the profitability of the firm. This may take the form of goodwill but, nonetheless, does contribute to the success of the firm, e.g., an investment in pollution control devices.

Some investments are made for the purpose of generating revenue, while others reduce present costs. In either case, the cost of investment must be related in some manner to the expected benefits. The cost involved is normally the acquisition cost of the asset, and this information is easily obtainable from suppliers. The benefits are much more difficult to obtain and usually involve estimates of future income.

The methods for relating costs and benefits range from the determination of an accounting rate of return to a procedure known as discounted cash flow. These methods are discussed in Chapter 15. The previous information was presented in order to illustrate the process by which decisions regarding capital expenditures are made. The elaborateness of the system will vary from firm to firm, but not the elements in the thought process.

Record Keeping

In order to facilitate the previously mentioned thought processes, accurate and current records of the physical properties must be maintained.

These records can be maintained through a manual system or the use of automated equipment, if the firm is large enough.

Irrespective of the type of system, the following information should be maintained on a ledger card for every fixed asset:

1. Acquisition cost including installation.
2. Predetermined useful life and salvage value.
3. Descriptive physical details together with location and date of first use.
4. Depreciation method to be employed (straight-line, sum-of-years'-digits, or double-declining balance).
5. The amount of the depreciation charge per year.
6. Date of any transfers or relocations, including any additional costs of reinstallation.
7. Any subsequent improvements or additional capitalizable costs.
8. Year of retirement.
9. Salvage value actually received.

An immediate benefit of maintaining the preceding information is the ability to determine the amount of yearly depreciation from the ledger cards. As improvements or additions are made to the asset, the ledger cards are changed to reflect the new depreciation. The expense for the year is the total of the precalculated amounts of depreciation shown on the ledger cards for all the assets. Thus the depreciation for an asset may be calculated at the time of acquisition and is not recalculated at the end of every year.

In addition to the plant ledger cards, other records are maintained so that fixed assets are not indiscriminately purchased. These additional records include:

1. Serially numbered authorizations for capital expenditures. There should be a limited number of people authorized to approve capital expenditures.
2. General ledger control accounts for all major classes of assets.
3. Plant ledger accounts for the assets of a multiplant firm.

In summary, a system for fixed assets requires a large volume of data to safeguard the assets and prevent indiscriminate purchases. In addition, these details can serve as the data for future investment decisions. Past experiences with similar types of assets should serve as the model for future decisions.

SUMMARY

In manufacturing and retailing firms, assets which are classified as non-current may represent a substantial part of total assets. These assets include fixed assets, intangibles, and natural resources. The cost of noncurrent assets includes all expenditures necessary to render the asset suitable for its intended use and which are expected to benefit future periods. The cost is allocated over the expected useful life. The allocation process for fixed assets is called *depreciation*. For intangibles, the *allocation*, or *amortization*, is over the shorter of the legal or economic life, in no case exceeding 40 years. The allocation of the cost of natural resources is called *depletion* and is based on a units of output method.

Because of the magnitude of investments in noncurrent assets, management must have information for choosing among investment opportunities. In addition, users of the financial statements need to be informed of the nature of the allocation methods, the amount of the periodic charge, and the unallocated cost of the noncurrent assets. This is accomplished through the financial statements and accompanying footnotes.

QUESTIONS

9-1. (a) Define fixed assets.
   (b) How does this definition differentiate fixed assets from inventory?

9-2. Why are items, such as an abandoned plant or land held for future use, which possess the characteristics of fixed assets, not listed as fixed assets on a balance sheet?

9-3. What guidelines do accountants use in determining whether or not an expenditure relates to a particular asset?

9-4. State which of the following would be considered a part of the cost of a machine:
   (a) Purchase price.
   (b) Transportation charges.
   (c) Rewiring room where machine is to be installed.
   (d) Cost of advertising the fact that machine will produce superior output.
   (e) Cost of training employees to operate machine.

9-5. Aaron Company purchased a tract of land and a building for $68,400. The land had an appraised value of $20,000 and the building was appraised at $70,000. How should the $68,400 cost be allocated to the land and building?

9-6. State which of the following would be considered a part of the cost of a building:
   (a) Grading and leveling the land.

(b) Architect fees.
(c) Cost of materials.
(d) Plant superintendent's salary while supervising construction.
(e) Taxes on the land during construction.

9-7. Define expenses and capitalizable costs as they apply to after acquisition expenditures on fixed assets.

9-8. Capitalizable costs can be accounted for in two different ways based upon the nature of the expenditure. Explain both methods indicating a situation in which each would be used.

9-9. Indicate whether the following items would be considered expenses or capitalizable costs. If the item is to be capitalized, state how the accounting would be handled:
(a) Buying new tires for delivery equipment.
(b) New addition to factory building.
(c) Cost of painting factory offices.
(d) Cost of lubricating and inspecting all factory machines.
(e) Replaced worn out machine engine with a new engine similar to the old one.
(f) Replaced wooden bed on a flat bed truck with a new steel bed.

9-10. Define and explain the nature of depreciation.

9-11. What assumption does each of the following depreciation methods make about the use of an asset over time?
(a) Straight-line
(b) Units of production
(c) Sum-of-years'-digits
(d) Declining balance

9-12. Which depreciation method would be best suited for a spare machine used periodically during the year when demand for the company's product increased.

9-13. Distinguish between the two accelerated depreciation methods, i.e., sum-of-years'-digits and declining balance.

9-14. (a) Oxford Company recently sold an asset which cost $50,000, and is ³⁄₄ depreciated, for $11,000. What entry should be made?
(b) Assume the asset in (a) was traded in along with $40,000 cash on a new asset having a list price of $60,000. What entry should be made using the list price method?

9-15. What are intangible assets? How do they differ from fixed assets?

9-16. Define amortization. Explain how it differs from depreciation.

9-17. Why is the units of production basis normally used in depleting natural resources?

9-18. What information should a company maintain with respect to its fixed assets?

9-19. What records are maintained by a company to prevent indiscriminate purchasing practices with respect to fixed assets?

PROBLEMS

Procedural Problems

9-1. In each of the following independent situations determine the cost of the assets involved and prepare journal entries for their purchase. Assume none of the assets will be productive until all costs associated with them are incurred.

   (a) Purchased a used machine at a cost of $15,000. Subsequent to the purchase the following expenditures were made:

   (1) General overhaul prior to use, $1,500.
   (2) Cost of course taken by employee to learn how to operate machine, $500.
   (3) Removal of a wall to make room for machine, $600.
   (4) Cost of insurance for machine for one year, $150.
   (5) Installation of air conditioning system for proper functioning of machine, $1,000.
   (6) New lighting system installed throughout the factory, $20,000. (The machine will occupy a space equal to 1/40 of the total factory.)

   (b) A tract of land, a building, and machinery were purchased for a lump sum price of $90,000. Legal fees for the purchase amounted to $1,800. On the day of the purchase, the three assets were appraised as follows: land, $60,000; building, $40,000; and machinery, $20,000.

9-2. The Commodore Corporation purchased a machine January 1, 1974, for $60,000. The machine is expected to have a ten-year useful life or 10,000 hours and a residual value of $5,000. Compute depreciation for the first and second year of use by the following methods:

   (a) Straight-line
   (b) Units of production (1,100 hours the first year, 1,500 hours the second year)
   (c) Sum-of-years'-digits
   (d) Double-declining balance

9-3. On March 31, 1974, the Volunteer Company purchased a new machine. The machine has an invoice price of $15,500 and a salvage value of $1,500. Transportation charges on the machine amounted to $800. The company was also required to provide special wiring for the machine at a cost of $300. Set up costs involved in making the machine operable amounted to $200. The accountant for Volunteer Company estimates the useful life of of the machine to be nine years or 300,000 units of output. The machine produced 10,000 units in 1974; 41,000 in 1975; and 26,000 in 1976.

   REQUIRED

   (a) Compute depreciation for the years 1974, 1975, and 1976 using the following methods:

   (1) Straight-line
   (2) Units of production
   (3) Sum-of-years'-digits
   (4) Double-declining balance

(b)   State which method you believe to be best and why.

9-4.   On January 3, 1958 the Red Tide Company purchased a new warehouse at a cost of $25,000. The warehouse had an estimated useful life of 20 years and a salvage value of $3,000. In January, 1968, the warehouse was completely remodeled at a cost of $8,000. The remodeling made the warehouse more suitable to the needs of the company, but it did not extend its useful life or increase its salvage value. On June 29, 1973 the warehouse was destroyed by fire. The warehouse was insured and the insurance company issued a check in the amount of $8,000 in full settlement.

REQUIRED.   Prepare journal entries for the following events:
(a)   Purchase of warehouse.
(b)   First year's depreciation (straight-line method).
(c)   Remodeling of warehouse.
(d)   Depreciation expense for the year 1968 (straight-line method).
(e)   Destruction of the warehouse and insurance settlement.

9-5.   Rebel Corporation purchased a new drill press on January 1, 1970 for $9,000. The drill press had a useful life of five years and a salvage value of $600. The corporation records straight-line depreciation on December 31 for each asset on hand at that date, and on the date of disposal for any asset traded or sold during the year. On May 1, 1973 Rebel Corporation traded the drill press on a new model. The new machine had a list price of $11,500, and Rebel Corporation was required to make a $7,500 cash payment in obtaining the new machine. The old machine had a secondhand market value of $3,600.

REQUIRED.   Record the asset trade using the following three methods:
(a)   List price method.
(b)   Tax method.
(c)   Secondhand market value method.
Which method gives the most accurate reflection of the true effect of this transaction in terms of value given up and value received?

9-6.   The SEC Company purchased a patent on April 1, 1974 for a total cost of $13,200. The patent had a legal life of 17 years and an economic life of 10 years. On January 5, 1978, the patent became worthless when a competing firm patented a similar, but more advanced, process.

REQUIRED.   Prepare journal entries for the purchase of the patent, the amortization for the first and second year, and the patent write-off.

9-7.   The Lotta Enterprise Company recently acquired a franchise to operate an Admiral Farraguts' Maine Boiled Lobster Restaurant. In securing the franchise at a cost of $22,000, plus $1,200 in legal fees, the company acquired sole rights to open these restaurants in the state of Kentucky. The franchise was to be amortized over the maximum allowable period, 40 years. The first restaurant was opened on January 1, 1964. For the period 1964 through 1968, the restaurant earned a modest profit, but much less than the company had anticipated. On January 2, 1969, it was decided that the franchise would be of value for a total of 20 years rather than the original estimate of 40 years. At this time, the amortization based upon the new estimate was put into effect. The restaurant began losing money, and on January 1, 1974, the franchise was considered worthless and written off. The restaurant was then converted to a fried chicken establishment.

REQUIRED. Prepare the journal entries for the purchase of the franchise, the first year's amortization, amortization for the year 1969, and the ultimate write-off of the franchise in 1974.

9-8. The following information pertains to three machines owned by Gator Company. The company records depreciation on December 31 each year using different methods for different machines.

Machine	Date purchased	Cost	Useful life	Salvage value	Depreciation method	Disposal date and details
M624	1/5/74	$6,500	12,000 units of product	$500	Units of production[a]	Sold for $4,000 on 5/21/75
M713	4/30/70	$3,500	5 years	$500	Straight-line	Traded on 1/2/72 for asset M823 trade-in allowance $2,000
M823	1/2/72	$4,300 less trade-in	5 years	$600	Sum-of-years'-digits	Sold 1/5/77 for $800

[a]Asset M624 produced 2500 units in 1974 and 500 units before its sale in 1975.

REQUIRED. Prepare journal entries to record:
(a) Acquisition of each machine.
(b) The yearly depreciation on each machine.
(c) The disposal of each machine. Use the tax method for the trade. (Only one entry is needed to record the disposal of machine M713 and acquisition of machine M823.)

Conceptual Problems

9-9. The Bulldog Company recently erected an addition to the plant. The addition was built by company employees who were relieved of their normal jobs to undertake the project. Although production decreased somewhat during the period of construction, the company believed the method would provide a substantial savings. The accountant for Bulldog Company is now attempting to compute the cost of the addition and asks for your help.

Materials used in construction cost $50,000, salaries of employees constructing the addition amounted to $20,000, the building permit cost $800, insurance on the structure during construction amounted to $500, and $200 was paid to an advertising firm for promotional services in connection with the opening of the new addition. The plant was operating at full capacity when construction began. General factory overhead costs (depreciation, heat, light, power, etc.) have consistently been $60,000 per year. Because of the construction activities, overhead increased to $80,000. One-fifth (1/5) of the original plant and employees were utilized in construction.

REQUIRED. Determine the capitalizable cost for the new addition.

9-10. The Callan Corporation was organized and all the stock was sold during 1972. The company executives immediately began purchasing equipment and hiring employees so that production could start on January 1, 1973. A building was purchased at a cost of $50,000 and was estimated to have a 15 year useful life and a salvage value of $5,000. Four identical machines were acquired; each machine cost $10,000, had a useful life of nine years and a salvage value of $1,000. A delivery truck was also purchased for $9,000. The truck is expected to have a useful life of five years and no salvage value.

In projecting earnings for the first five years of operations, the budgeting department estimated the following net income after expenses but before depreciation and taxes: 1973–$16,000; 1974–$18,000; 1975–$20,000; 1976–$26,000; 1977–$40,000. The budgeting department also reported that the company would be in a rather tight liquidity position in its first three years of operations.

A difference of opinion existed in the attempt to decide upon a depreciation method for the company's assets. The chief finance officer believed the straight-line method to be best, while the head of accounting favored the sum-of-years'-digits method. You have been asked to give your opinion as to the most appropriate method assuming the foregoing estimates are valid. Included in your analysis should be a schedule of net income after depreciation and taxes for the years 1973–1977. Assume a tax rate of 50%. Justify your position.

9-11. The Klondike Company purchased a building on January 1, 1972 for $120,000. The building was estimated to have a useful life of 40 years and a salvage value of $10,000. The company uses straight-line depreciation for all fixed assets and the building was depreciated in that manner for the years 1972 through 1974. In January 1975 a new chief accountant was hired to replace the former chief accountant who had retired. In reviewing the fixed assets and depreciation policy, the new accountant decided that the useful life of 40 years was unrealistic in terms of the type of construction used in the building. The new chief accountant thus revised the useful life of the building to 60 years in order to more nearly reflect its economic life. The problems he now faces are (a) how to reallocate depreciation charges in future years and (b) whether or not the reported net income in the years 1972 through 1974 should be restated.

REQUIRED
(a) Determine the effect on net income in the years 1972–1974 of having depreciation based upon 40 years rather than 60 years.
(b) Show the amount of depreciation which will be recorded in the years subsequent to 1974.

9-12. The following assets are to be used by the Kieso Company in its operations which will begin shortly. The company is interested in using a depreciation method for each type of asset which will closely approximate the assets' actual deterioration:

1. Building. Purchase price was $125,000; $500 was spent in legal fees and title search. The roof had to be repaired at a cost of $2,500, and new facilities were constructed at a cost of $2,000. The estimated useful life is 25 years with a $5,000 salvage value.

2. Machinery. Three machines costing $10,000 each were purchased new. Freight and installation costs amounted to $700 per machine. The machines have a useful life of five years if utilized to produce their capacity of 50,000 units per year. Salvage value is $500 per machine.

3. Equipment. A used delivery truck was purchased for $7,000. The engine on the truck was overhauled at a cost of $800, making the truck usable for 100,000 miles. The company plans to purchase a new truck at the end of six years. There will be no salvage value. Six typewriters were purchased for a total of $1,650. The typewriters can be used for seven and one-half years, but the manufacturer estimates that new innovations in the industry will cause them to be obsolete in four years. Salvage value totals $150.

REQUIRED. Determine the depreciation method to be used on each type of asset so as to comply with the previously stated policy of the company. Give the reasons for your choices. Also, using the same method you just used, compute the depreciable cost of the asset. (Explain any assumptions made.)

10
LIABILITIES

EDUCATIONAL OBJECTIVES

The material in this chapter is designed to achieve several educational objectives. These include:

1. An understanding that the purpose of the equity side of the balance sheet is to reflect the various sources of the assets.
2. An insight into the factors that should be considered in determining the differences between the debt and equity classification.
3. An awareness of the criteria used for distinguishing current liabilities from long-term liabilities.
4. An understanding of the various types of bonds and the nature of the bond contract, or indenture.
5. A knowledge of the reason for bonds being sold at a discount or premium and the accounting procedures necessary to record bond transactions.

The purpose of the preceding five chapters was to present information about the assets generally reflected on balance sheets of business organizations. Each asset or asset category has certain unique characteristics; however, they all have one common characteristic. In order for any tangible or intangible item to be classified as an asset, it must be capable of providing some benefit to the business beyond the current period. In accounting language, this expected benefit is defined as *future service potential*.

In this chapter and the following one, attention will be directed to the equity section of the balance sheet. This chapter is concerned with the liabilities. Chapter 11 will present information about the owners' equity portion of the balance sheet.

One of the major problems in accounting for equities is that equities have no unique characteristics as do assets. In fact, there are often conflicting views among accountants, as well as among the users of financial statements, regarding the existence or nonexistence of certain liabilities.

In Chapter 3, emphasis was given to the equality of the basic accounting equation — Assets = Equities. The equation could have been expressed as Resources = Sources. This means that the assets (future service potential) of the firm are generated from particular sources (equities). These sources may include purchases on credit, borrowing, investment by the owners, and earnings retained in the business.

The purpose of the equity side of a balance sheet is to provide a summary of the sources and claims on the assets. Such a summary is useful for several reasons. One is that, legally, certain claims have higher priorities than other claims. Secondly, owners and creditors are interested in their relative percentage claim on assets. Another reason is that creditors are especially concerned that the owners are bearing the major risk in the operation of the business.

There are numerous fine legal points distinguishing the various levels of claims; however, there are at least five major categories of claims which can best be viewed from the point of view of their legal priority in the event of liquidation:

1. The highest priority is given to wages due employees and taxes owed governmental units.

2. The second priority claims are those which are secured by specific assets of the business, such as a mortgage loan protected by a claim on a building.

3. The third level includes those claims which are unsecured and, therefore, protected only by the over-all financial stability of the business.

4. The fourth level priority is given to the claims of preferred stock-

holders. Preferred stock is an ownership security which gives the holder some specified preference over common stockholders.

5. The final claim is that of the common stockholders. These claimants receive whatever remains after all other claims have been satisfied. If the business is successful, their claims may be considerably more than their original investment. On the other hand, if the business has not been a success, there may be no return to these claimants.

The first three levels of claims are typically reflected as liabilities on the balance sheet while the last two are shown as owners' equity.

There are, however, certain situations in which it is necessary to rely upon factors other than the level of priority in order to determine the appropriate balance sheet presentation. These situations involve securities which cannot be clearly labeled as a liability or owners' equity. The Internal Revenue Service has a special interest in this matter, because payments for the use of borrowed funds are deductible by the firm for tax purposes as interest. On the other hand, distributions of profits to the owners are referred to as *dividends* and are not deductible for tax purposes.

Criteria for Debt-Equity Classification

The following factors influence the debt or equity classification of particular securities. These criteria have evolved from court decisions, involving the Internal Revenue Service, and writings in accounting, finance, and law:

1. Claim on assets
2. Maturity date
3. Maturity value
4. Claim on income
5. Right to enforce payment
6. Voice in management
7. Business reasons for issuing security
8. Relationship of owner to creditor

The weight to be given each of these factors in determining whether a security represents debt or equity is not clear. In some cases the courts have given consideration to one factor to the exclusion of others. Therefore, the distinction between debt and equity is not subject to a generalized definition.

CLASSIFICATION OF LIABILITIES

A distinction is made on the balance sheet between current and long-term liabilities similar to the classification of assets. In the case of assets, the purpose was to distinguish between those to be used in the current operation of the business and those expected to benefit the firm over a longer period of time. In the case of liabilities, the purpose is to draw attention to those obligations of the business likely to require the use of funds within the next accounting period, and those which will not require payment for some extended period of time.

Current Liabilities

The American Institute of Certified Public Accountants defines a *current liability* as follows:

> The term current liabilities is used principally to designate obligations whose liquidation is reasonably expected to require the use of existing resources properly classifiable as current assets or the creation of other current liabilities. As a balance sheet category, the classification is intended to include obligations for items which have entered into the operating cycle, such as payables incurred in the acquisition of materials and supplies to be used in the production of goods or in providing services to be offered for sale; collections received in advance of the delivery of goods or performances of services; and debts which arise from operations directly related to the operating cycle, such as accrual for wages, salaries, commissions, rental, royalties, and income and other taxes. [AICPA, *Accounting Research and Terminology Bulletins,* Final Edition (New York: 1961), pp. 21–22]

The *operating cycle* referred to in the definition is the time required to buy or manufacture a product, sell it, convert the receivable into cash, and buy new merchandise or manufacture new products (see Figure 10-1). Obviously, the length of this cycle will vary from one business to another. Therefore, the classification of current liabilities may differ among companies. In practice, the definition has been modified so that the criterion for classification as a current liability is the operating cycle or one year, whichever is longer. In many industries there are several operating cycles within a year, e.g., a grocery store. When there is more than one cycle in a year, the one year criterion is used. In other industries, e.g., tobacco, liquor, etc., the operating cycle may extend over several years. In that case, the current liabilities are those obligations which will be paid within the operating cycle.

Among the common current liabilities are: trade accounts payable, notes payable, unearned collections, and various accrued items such as salary payable, interest payable, and taxes payable.

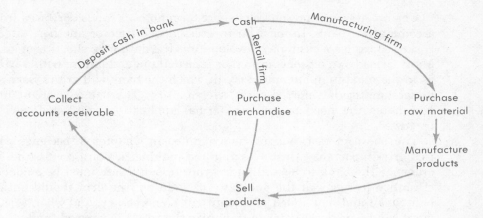

FIGURE 10-1

Operating cycle

Trade Accounts Payable. This account reflects the cost of materials and supplies acquired for use in the normal operation of the business and for which payment has been postponed. In Chapter 3, this account was presented in the context of a purchases journal; therefore, the discussion will not be repeated.

Notes Payable. For a note payable to be classified as a current liability, it must be relatively short-term. Such notes, typically, originate from one of two sources. A supplier of materials may have extended credit beyond the time normally allowed for accounts payable. This may occur because of the buyer's inability to pay during the regular period. In that case, the recording of the note would take the form:

Accounts Payable	500	
Notes Payable		500

The conversion from Accounts Payable to Notes Payable does not insure the seller of receiving payment, but the note does provide written evidence of the promise to pay. In addition, the note payable will probably have a stated interest rate which the firm will earn by extending payment.

According to Accounting Principles Board Opinion Number 21, in those cases in which a note is issued without a stated rate of interest there is a presumption that a fair rate of interest was included in the face amount. Thus the note must be recorded at the present value of the amount to be received at maturity. The difference between the present value and the face value of the note payable is a discount, which is amortized over the life of the note.

The second common source of short-term notes is funds borrowed from a commercial bank. In most businesses, at one time or another, a temporary shortage of cash may develop, thus the need for a short-term loan. Many firms have either a formal or informal understanding with a bank as to the maximum amount of credit which the bank will grant. Such an understanding is referred to as a *credit line*. The use of a credit line eliminates the need to make a formal application each time cash is borrowed.

The use of a cash budget, as explained in Chapter 5, becomes particularly important in establishing and making use of a credit line. In order for the bank to be willing to grant a loan, there must be evidence of ability to repay at the due date. In addition, the firm should anticipate when and how much of a loan will be necessary, and when repayment can be made in order to minimize the cost of borrowed funds.

Unearned receipts. In certain industries, a common practice is the receipt of a payment for a service or product prior to the performance of the service or delivery of the product. An example of this is publishing companies. In this industry, payment for subscriptions may be received for one or more years in advance. The company has an increase in assets as a result of receiving cash, but it would be inappropriate to record the total amount as earned revenue. The unearned portion of the receipt is recorded as a liability, since the company has not completed its obligation under the contract.

When the cash is received, it may be recorded in any one of three ways. For example, assume that on July 1, 1974, $500 is received for one-year subscriptions to a monthly publication. The collection may be recorded (a) one part as earned revenue and one part as unearned revenue, (b) all as earned revenue, or (c) all as unearned revenue.

The first method would eliminate the need for an adjusting entry at the end of the first year, because the amount recorded as earned revenue was the amount earned.

Cash	500	
Unearned Revenue		250
Earned Revenue		250

If the second method is used, the adjusting entry must transfer part of the subscription from Earned to Unearned Revenue:
Initial entry:

Cash	500	
Earned Revenue		500

Adjusting entry:

Earned Revenue	250	
Unearned Revenue		250

Under the third method the entries would be:
Initial entry:

Cash	500	
Unearned Revenue		500

Adjusting entry:

Unearned Revenue	250	
Earned Revenue		250

Note that in all three methods, after the adjusting entry, the amount of earned revenue for 1974 is $250. Also, the current liability at December 31, 1974 under all three methods is the same.

Accrued Liabilities. A substantial part of the adjusting process at the end of an accounting period involves the accrual of current liabilities. Such adjustments are necessary because of certain economic events which have not yet resulted in transactions and, therefore, have not been recorded. These events must be recorded in order to properly reflect the expenses in the income statement as well as the liabilities on the balance sheet.

Salary payable. Unless employees are paid each day, it is likely that one or more days wages will have been earned, but not paid, at the end of the accounting period. Consequently, it is necessary to calculate, and accrue, the salaries which have been earned but not paid. The entry to record the accrual of unpaid salaries of $800 is:

Salary Expense	800	
Accrued Salary		
Payable		800

The illustration reflects only a nominal amount of salary cost and accounting procedures; however, payroll procedures normally involve substantial technical details. Many of these technical procedures can be effectively performed by a computer. Rates of pay, federal income withholding rates, number of deductions, state withholding rates, social security rates, and any other pertinent information can be included in the computer program for the payroll. In order to calculate the actual payroll or the accrual at the end of the year, the only input required would be the employee's name and number of hours worked.

Interest payable. Most loans require the payment of interest at a specified rate and time. If the payment date does not coincide with the end of the accounting period, an adjustment is necessary. The amount of the adjustment is the accrued interest since the last interest payment was made or the date the loan was obtained.

The classification of Interest Payable as a current liability is not dependent upon whether the loan is current or long-term. As with all current liabilities, the criterion is the expected payment date; therefore, if the interest will be paid during the next accounting period, it is a current liability.

The interest to be paid on long-term liabilities is recorded as an expense and payable only for the current period. The total interest to be paid over the life of the loan is not recorded as an expense of the period and a liability at the time the loan is obtained. Interest is a cost of using money for a period of time; thus there is no expense until the firm has had the use of the money. In order to record interest payable for the life of the loan, it would be necessary to record a debit to Interest Expense or Prepaid Interest. Neither alternative would properly reflect the financial status of the firm because it is not an expense of the period, nor is the interest prepaid.

Taxes payable. There are several taxes which must be accrued at the end of the period in order to reflect the appropriate expense and liability. Certain taxes must be estimated since they will not have been assessed at the end of the fiscal period. The more common accounts involving tax liabilities are: Income Taxes Payable; Payroll Taxes Payable; Property Taxes Payable; and Sales Taxes Payable.

The amount of income taxes payable shown on the balance sheet is the estimated amount of taxes to be paid to the federal government. Payroll Taxes Payable are the taxes to be paid to the local, state, and federal agencies for the taxes deducted from the employees' wages, as well as the employers' share of the taxes. Property Taxes Payable represent an estimate of the amount of taxes which have been accrued since the previous tax payment. Sales Taxes Payable represent the amount of sales tax collected from customers and not yet remitted to the taxing authority.

LONG-TERM LIABILITIES

Current liabilities relate primarily to those items involved in the day-to-day operations of the business. Long-term liabilities, on the other hand, reflect the interest of long-term creditors in the operation of the business.

Among the more significant long-term liabilities likely to be found on a balance sheet are:

1. Bonds payable
2. Notes payable
3. Mortgages payable
4. Long-term lease obligations
5. Deferred tax liability

The first four items—bonds, notes, mortgages, and leases—represent methods of obtaining resources through long-term borrowing. Deferred tax liability is the result of differences between accounting practices and income tax regulations. Each of these items will be explained in greater detail in the remainder of this chapter.

Bonds Payable

Bonds are a common source of borrowing long-term funds. Governmental units, nonprofit organizations, and businesses make extensive use of the bond market as a source of capital.

Each bond issue contains an agreement between the bondholders and the company. The document which sets forth the agreement is called a *bond indenture*. The possible terms in bond agreements seem to be limited only by the ingenuity of the person drafting the provisions. Therefore, it becomes important, not only for accounting purposes but also for investment decisions, to study carefully the conditions contained in the bond agreement.

Even though there are many differences, there are certain points common to all bond indentures:

1. The amount of the bond. This is referred to as *face value* since this value is printed on the face of the certificate. Bonds are usually sold in multiples of $1,000.
2. The *annual rate of interest* to be paid.
3. The date on which the bonds are to be repaid. This date is called the *maturity date*.

In addition, bond agreements often include one or more of the following provisions:

1. The bonds may be *secured* by specific assets of the firm, e.g., buildings or machinery. Thus the bondholders have a prior claim on the pledged assets in the event of liquidation. Obviously, this feature is unimportant when the business is operating profitably. In addition, bondholders loan funds based on the expectation of being repaid out of earnings, not out of liquidation proceeds.
2. If the bonds are not secured by specific assets, they are called *debenture bonds*. Such bonds rank with all other unsecured claims in the event of liquidation.
3. The bonds may be *convertible*. This means that given certain conditions, usually relative to market price, the bonds may be converted into shares of common stock of the issuing company. The potential for conversion into common stock is reflected by the market price of these bonds rising and falling with the market price of the common stock. There has been substantial disagreement over the appropriate balance sheet classification of convertible bonds.

Many individuals believe that some portion, if not all, of a convertible bond issue should be recorded as owners' equity rather than as debt. The present generally accepted accounting position is that these bonds should be shown as debt until actual conversion occurs.

4. In order to give the company some flexibility relative to the timing of repayment, bonds may be *callable*. This means that the bond indenture contains a provision which states that, *at its option*, the company may announce that the bonds are to be returned to the company for payment. One reason for calling bonds may be that the current market rate of interest is lower than the stated rate on the bonds outstanding. In many cases, in order to retire the old bonds, the company may sell new bonds at the lower rate to obtain the necessary funds. This process is called *bond refunding*.

5. The indenture may state that a certain portion will be redeemed each year rather than the entire amount being repaid at maturity. These bonds are called *serial bonds*. The amount of the serial bond to be redeemed during the next year is classified as a current liability on the balance sheet. The remaining portion is a long-term liability.

Accounting for Bond Issues.   Bonds, sold at face value, do not present any difficulty in making a journal entry. There is an increase in the asset received and an increase in a corresponding liability. For example:

Cash	1,000	
Bonds Payable		1,000

The entry for the payment of interest for a year, assuming a 6% bond, would be:

Interest Expense	60	
Cash		60

The preceding simple illustration is seldom found in actual practice. As previously discussed, bond indentures specify a face value and interest rate. The interest rate must be determined some time prior to the issue date since the indentures must be printed. The rate is set on the basis of what management believes to be necessary in order for the bonds to sell on the market; however, the rate of interest which investors are willing to accept is constantly changing. For a review of the process by which the market price is established see Chapter 8. Although that discussion was from the point of view of the investor, the same process is applicable to the issuing firm.

As discussed in Chapter 8, bond prices are stated as a percentage of face value rather than as a dollar amount. When the market rate of interest is higher than the stated rate and the bonds are sold at a price below face value, the bonds are said to be sold at a *discount*. If the market rate of interest is lower than the stated rate, the bonds will be sold at a price above the face value and the bonds are said to be sold at a *premium*. Thus, the amount of discount or premium is the difference between the selling price and the face value.

Bonds sold at a discount.   The accounting procedures for bonds issued at a discount will be explained and illustrated assuming the following facts:

Face value of bonds	$500,000
Stated rate of interest	6%
Maturity date	January 1, 1984
Date bonds sold	January 1, 1974
Actual selling price (95% of face value)	$475,000
Interest payment date	June 30 and December 31

On January 1, 1974, the entry for the sale is:

Cash	475,000	
Bond Discount	25,000	
Bonds Payable		500,000

Note that only $475,000 of cash was received but a liability of $500,000 was created. The difference of $25,000 is recorded as a debit to Bond Discount because, in accounting, liabilities are recorded at their face value. The preferred accounting treatment of the Bond Discount account is to reflect it as an offset to Bonds Payable on the balance sheet, resulting in a net liability of $475,000. This method is preferred because the net liability represents the present value of the face amount of the bonds.

On January 1, 1984, when the bonds mature, the company must pay the bondholders the full $500,000 face value even though only $475,000 was received when the bonds were sold. Thus the company is paying for the use of the borrowed funds in two ways. There is an annual interest payment of $30,000 (6% × $500,000) and at maturity a payment of $25,000 more than originally received. Therefore, the total cost of the borrowed funds over the life of the bonds is:

Maturity payment	$500,000
Interest (10 years × $30,000)	300,000
Total payments	$800,000
Proceeds from issue	475,000
Total cost	$325,000

The total cost is increased by the $25,000 discount; therefore, the discount is effectively an increase in interest expense and should be allocated over the life of the bonds even though it will not be paid until the maturity date of the bonds. The amortization may be on a straight-line or present value basis. Since the present value method prorates the discount or premium over the life of the bonds on the basis of a constant rate applied to the carrying value, it is more theoretically correct than the straight-line basis. The straight-line method, however, adequately demonstrates the conceptual aspect of amortization and will be used in the following illustration. The entry to be made December 31 of each year is:

Interest Expense	17,500	
Bond Discount		2,500
($25,000 × 1/10)		
Cash		15,000
($30,000 × 1/2 year)		

The credit to Bond Discount ($2,500) is the annual amortization of the discount on a straight-line basis, which increases interest expense. Since the Bond Discount account is reduced by $2,500 each year, at the end of ten years this account will have a zero balance and the net liability for bonds payable will be $500,000. Thus, the balance in the liability account is equal to the amount of the repayment.

The following schedule reflects the situation relative to the bonds issued at a discount:

Year	(A) Amount of interest actually paid	(B) Amount of discount allocated	(A) + (B) Total interest expense	(C) Amount of discount unallocated	(D) Bonds payable	(D) − (C) Net long-term liability as of 12/31
1974	$ 30,000	$ 2,500	$ 32,500	$22,500	$500,000	$477,500
1975	30,000	2,500	32,500	20,000	500,000	480,000
1976	30,000	2,500	32,500	17,500	500,000	482,500
1977	30,000	2,500	32,500	15,000	500,000	485,000
1978	30,000	2,500	32,500	12 500	500,000	487,500
1979	30,000	2,500	32,500	10,000	500,000	490,000
1980	30,000	2,500	32,500	7,500	500,000	492,500
1981	30,000	2,500	32,500	5,000	500,000	495,000
1982	30,000	2,500	32,500	2,500	500,000	497,500
1983	30,000	2,500	32,500	—0—	500,000	500,000
	$300,000	$25,000	$325,000			

Bonds sold at a premium. In the discussion of bonds sold at a premium, the facts are assumed to be as follows:

Face value of bonds	$500,000
Stated rate of interest	6%
Maturity date	January 1, 1984
Date bonds sold	January 1, 1974
Actual selling price (106% of face value)	$530,000
Interest payment date	June 30 and December 31

The entry to record the sale on January 1, 1974 is:

Cash	530,000	
Bonds Premium		30,000
Bonds Payable		500,000

In this case, the journal entry reflects the fact that cash of $530,000 was received for a bond payable of $500,000. The bond premium is added to the Bonds Payable account on the balance sheet. Thus the net liability at January 1, 1974 is $530,000, even though only $500,000 will be repaid. The calculation of total cost of the bonds is:

Maturity payment	$500,000
Interest (10 years × $30,000)	300,000
Total payments	$800,000
Proceeds from issue	530,000
Total cost	$270,000

Note that the total cost is decreased by the $30,000 premium. Therefore, the bond premium is effectively a decrease in interest expense and should be allocated over the life of the bonds even though it has already been collected. The entry to be made on December 31 of each year is:

Interest Expense	12,000	
Bond Premium	3,000	
($30,000 × 1/10)		
Cash ($30,000 × 1/2 year)		15,000

The debit to Bond Premium is the annual amortization and reduces that account by $3,000. At the end of ten years, this account will have a zero balance and the net liability for bonds payable will be $500,000.

The following schedule reflects the situation relative to the bonds issued at a premium:

Year	(A) Amount of interest actually paid	(B) Amount of premium allocated	(A) − (B) Total interest expense	(C) Amount of premium unallocated	(D) Bonds payable	(D) + (C) Net long-term liability as of 12/31
1974	$ 30,000	$ 3,000	$ 27,000	$27,000	$500,000	$527,000
1975	30,000	3,000	27,000	24,000	500,000	524,000
1976	30,000	3,000	27,000	21,000	500,000	521,000
1977	30,000	3,000	27,000	18,000	500,000	518,000
1978	30,000	3,000	27,000	15,000	500,000	515,000
1979	30,000	3,000	27,000	12,000	500,000	512,000
1980	30,000	3,000	27,000	9,000	500,000	509,000
1981	30,000	3,000	27,000	6,000	500,000	506,000
1982	30,000	3,000	27,000	3,000	500,000	503,000
1983	30,000	3,000	27,000	— 0 —	500,000	500,000
	$300,000	$30,000	$270,000			

The amortization of a bond discount for the issuing company is an increase in interest expense, whereas the amortization of a discount results in an increase in interest income for the investor. The amortization of a bond premium has the reverse effect; it decreases interest expense for the issuing company and decreases interest income for the investor.

Retirement of bonds. As previously indicated, by amortizing the premium or discount, the net liability for bonds payable will be equal to face value at the maturity date. Therefore, the entry to record retirement of the bonds in either of the two illustrations is:

Bonds Payable   500,000
     Cash   500,000

Bonds may be retired prior to the maturity date, in which case any unallocated premium or discount must be removed from the accounts. To illustrate the journal entry, assume that the bonds sold at a premium were retired at the end of 1980. As a penalty for retiring the bonds early, the company was required to pay 1% more than the par value. The entry for the retirement is:

Bonds Payable   500,000
Bonds Premium   9,000
     Cash   505,000
     Gain on Retirement
     of Bonds   4,000

Note that the penalty of 1%, or $5,000, reduced the gain from a potential $9,000 to $4,000. When bonds are retired the amortization of the bond premium or discount must be brought up-to-date so that the value of the bond can be determined. This example assumed that the entry for the amortization and interest payment had been made since the retirement occurred at the end of the year.

Notes Payable

Long-term notes payable have the same characteristics as short-term notes, except that the due date extends beyond one year. The notes are presented on the balance sheet at the net amount of the liability at year end.

Mortgages Payable

In the earlier discussion relative to the various levels of claims on assets in the event of liquidation, the second-order priority went to claims secured by specific assets. A claim of this type is usually supported by a mortgage. This is a document setting forth the agreement as to the amount and the specific assets pledged as security for the loan.

In addition to the individual mortgage holder, other creditors have an interest in knowing about the mortgaged property. The claims of other creditors can only be satisfied by assets other than those pledged as collateral for the mortgage. For this reason, all significant details concerning mortgages payable should be presented on the balance sheet. The information may be included in the body of the financial statements or in a footnote.

Long-Term Lease Obligations

In recent years, a relatively new industry has developed. The firms in this industry are concerned with the acquisition of various capital goods, such as buildings and equipment, for rental purposes. The firm making the purchase has no intention of using the goods in production; rather, the company expects to earn a profit from the excess of the rent received over costs.

When the agreement between the two firms involved is strictly a long-term rental, accountants do not record any asset or liability for the user. This position has been criticized on the grounds that the right to use the property over an extended period of time provides the future service potential necessary for asset recognition. Furthermore, the lease agreement obligates the firm to make payments for a specified period of time; thus it has characteristics of a liability. Even though the lease is not recorded, pertinent information about the lease should be included in a footnote to the financial statements.

Many lease arrangements are such that the user (lessee) of the goods is, in substance, the owner. In these situations, financing of the asset is the only service provided by the purchaser (lessor). Consequently, the accounting view is that this agreement is, *in fact*, a purchase and should be recorded as such. The property involved would be shown as an asset and the obligation for payments would be recorded as a liability.

The amount to be recorded as an asset and liability is the present value of the future lease payments discounted at an appropriate rate. To illustrate the recording of leases on the books of the lessee, assume that the lessee has a lease-purchase agreement with a discounted present value of $73,600. The entry would be:

Rights to Leased Property	73,600	
Liability under Lease Contract		73,600

The Rights to Leased Property account is shown on the balance sheet under the caption of Property, Plant, and Equipment and is amortized over the period of the lease.

In recent years, the accounting profession and the Securities and Exchange Commission have given considerable attention to the adequate disclosure of lease commitments. This has resulted because leasing is such an integral part of the operations of many firms and it is often reflected only in footnotes. Since footnotes are considered to be an integral part of the financial statements, a user of the financial statement must give careful attention to any information contained in the footnotes.

Deferred Tax Liability

The Internal Revenue Code was enacted by Congress for the purpose of generating the revenue considered necessary for the operation of the government. In addition, various economic goals were considered. These objectives do not necessarily correspond to the most appropriate accounting practices. Therefore, it is not unusual for some revenue or expense items to be included in one period for tax purposes and in another period for accounting purposes.

The income tax expense shown on the income statement for accounting purposes should relate to the accounting income. Since there are differences in the concepts of accounting and taxable income, the taxes paid may be different from the tax expense.

A common reason for a difference between accounting income and taxable income is the use of different depreciation methods. In order to illustrate the concept of deferred income tax, assume that Sharp Company had accounting and taxable income of $50,000 before deducting depreciation in each of four years. A fixed asset was purchased for $100,000 at the beginning of 1975 with no salvage value and a four-year life. For tax purposes, the sum-of-years'-digits method is used, while for book purposes straight-line is used:

Accounting income

	1975	1976	1977	1978
Net income before depreciation	$50,000	$50,000	$50,000	$50,000
Depreciation (straight-line)	25,000	25,000	25,000	25,000
Net income before taxes	$25,000	$25,000	$25,000	$25,000
Taxes (40%)	10,000	10,000	10,000	10,000
Net income	$15,000	$15,000	$15,000	$15,000

Taxable income

	1975	1976	1977	1978
Net income before depreciation	$50,000	$50,000	$50,000	$50,000
Depreciation (sum-of-years'-digits)	40,000	30,000	20,000	10,000
Net income before taxes	$10,000	$20,000	$30,000	$40,000
Taxes (40%)	4,000	8,000	12,000	16,000
Net income	$ 6,000	$12,000	$18,000	$24,000

To summarize:

Year	Tax expense	Tax payable	Deferred tax at year end
1975	$10,000	$ 4,000	$6,000
1976	10,000	8,000	8,000
1977	10,000	12,000	6,000
1978	10,000	16,000	—0—
	$40,000	$40,000	

The Deferred Tax Payable account is a record of the amount of the taxes which have been postponed as a result of using different methods for accounting and tax purposes. Since the taxes are postponed, not avoided, the tax will be eventually paid. The advantage to the company of postponing the payment of the taxes lies in the fact that the company has the use of the cash until the taxes must be paid.

Balance Sheet Presentation

This chapter has discussed several of the more common liabilities presented in financial statements. In order to illustrate the liability section, a balance sheet which contains all of the items discussed is presented.

XYZ Company
Balance Sheet
December 31, 1974

Liabilities

Current Liabilities

Trade accounts payable	$ 98,017	
Notes payable, due March 15, 1975	14,500	
Salaries payable	5,150	
Interest payable	5,187	
Taxes payable	4,300	
Unearned receipts	21,742	
Total Current Liabilities		$148,896

Long-Term Liabilities

Notes payable, 7% due December 31, 1977		$ 20,000	
Mortgage payable, 6%, 5 years, due 1979[a]		50,000	
Bonds payable, 7½% debentures due January 1, 2004	$200,000		
Less: Unamortized discount on bonds	7,500	192,500	
Long-term lease obligations[b]		145,265	
Deferred tax liability		33,083	
Total Long-Term Liabilities			440,848
TOTAL LIABILITIES			$589,744

[a]The mortgage payable is secured by printing equipment whose current market value is $60,000.
[b]The long-term obligation is on the home office building, which can be purchased at the end of the lease contract for $1.

SUMMARY

There are several factors which distinguish liabilities from owners' equity. These factors are: claim on assets, maturity date, maturity value, claim on income, right to enforce payment, voice in management, business reasons for issuing security, and relationship of owner to creditors. The distinction is important because of legal differences and the relative risk factors of owners and creditors.

Liabilities are classified as current and long-term on the balance sheet in order to draw attention to those which will require the use of funds within the next accounting period and those which will not require payment for some extended period of time. Among the common current liabilities are: trade accounts payable, notes payable, and salaries payable. In contrast, long-term liabilities are more likely to involve financing contracts such as bonds, leases, and mortgages.

QUESTIONS

10-1. (a) What are equities?
     (b) Name the two major categories of equities.
     (c) What is a major problem in accounting for equities?

10-2. In the event of liquidation, list the five major categories of priority claims and state how each is classified on the balance sheet.

10-3. (a) Why does the Internal Revenue Service have an interest in the classification of equities?
     (b) List eight factors which influence debt or equity classification.

10-4. (a) What are the two groups of liabilities on the balance sheet?
     (b) According to the American Institute of Certified Public Accountants, what is a current liability?
     (c) Define *operating cycle*.

10-5. (a) Name two sources from which notes payable originate.
     (b) Define *credit line*.

10-6. Distinguish between unearned receipts and accrued liabilities in terms of cash flow.

10-7. (a) Name four long-term liabilities which represent methods of obtaining resources through long-term borrowing.
     (b) How is deferred tax liability different from other long-term liabilities?

10-8. Define the following terms:
     (a) Bond indenture
     (b) Face value
     (c) Maturity date
     (d) Secured bonds
     (e) Debenture bonds
     (f) Convertible bonds
     (g) Callable bonds
     (h) Bond refunding
     (i) Serial bonds

10-9. (a) Why are the face rate and the market rate on bonds usually different?

(b) How is this resolved?

10-10. (a) When will bonds sell at a discount?

(b) When will bonds sell at a premium?

10-11. In addition to the individual mortgage holder, other creditors have an interest in knowing about mortgaged property. Why?

10-12. What is the accounting treatment of a long-term rental? Why has it been criticized?

10-13. The Pine Cone Corporation is issuing bonds with a par of $2,000,000 and an interest rate of 7½%. Similar bonds have been sold and a market rate of 4% exists. What will be the approximate selling price, assuming a perpetual existence?

10-14. On June 1, 1974 the Ballard Bat Company issued some bonds at 104. The bonds are dated October 1, 1973 and are four-year bonds. On December 3, 1974, the Ballard Bat Company amortized $700 of the premium. What is the par value of the bonds?

PROBLEMS

Procedural Problems

10-1. The following information is available for adjusting the books of Green Company at December 31, 1974:

(a) The firm issued ten-year, 7% bonds, par $200,000, on March 1, 1974 at par. Interest is payable on February 1 and August 1.

(b) Employees earn $5,000 per week and are paid each Friday. December 31, 1974 is a Monday.

(c) Net income on the company's books is estimated to be $500,000 while taxable income per the tax return is estimated to be only $400,000. The corporate income tax rate is 48%.

(d) The company borrowed $1,200 from the bank on November 16, 1974. A 60-day, 8% note payable was signed.

(e) The city collects property taxes for their fiscal year, July 1 to June 30. The tax bill is not mailed until January of each year with the payment due in March. The company estimates that its property tax for the 1974–1975 fiscal year will amount to $5,400.

(f) Miscellaneous selling expenses incurred but not yet paid amount to $150.

REQUIRED. Journalize the necessary adjusting entries on December 31, 1974.

10-2. The Rightly King Publishing Company has a situation involving unearned receipts as of December 31, 1974:

A new monthly magazine, *The Royale*, was first published in September, 1974. When cash is received for annual subscriptions in any one month, the first issue is mailed the following month. Cash received during 1974 was as follows:

August, $10,500
September,   10,200
October,   9,900
November,   9,600
December,   9,547

REQUIRED. Journalize the original entry, the adjustment on December 31, 1974, and the adjustment on December 31, 1975, assuming the revenue is recorded:

(a) A portion as earned revenue, the remainder as unearned revenue.
(b) All as earned revenue.
(c) All as unearned revenue.

10-3. On January 1, 1974, the Smolinski Corporation issued $100,000 of ten-year, 5% bonds at 95. On the same day, the Wilson Company issued $100,000 of ten-year, 6% bonds at 105. In both cases interest is payable on June 30 and December 31.

REQUIRED

(a) Journalize the entries related to the bonds for 1974 for Smolinski Corporation.
(b) Journalize the entries related to the bonds for 1974 for Wilson Company.
(c) Compute bond interest expense for 1974 for each company.

10-4. The Art Company issued $4,000,000, 6% bonds, at 104 on September 1, 1973. The bonds are dated May 1 and interest is payable on May 1 and November 1. The bonds mature on May 1, 1980.

REQUIRED. Prepare all journal entries concerning the bonds for the years 1973, 1974, and 1980. Entries should include sale of the bonds, semiannual interest payments, year-end interest accrual and premium amortization, and redemption of the bonds.

10-5. Farmer Publishing leased a warehouse to the Queen Bindery, collecting one year's rent of $1,800 on November 1, 1974.

REQUIRED. Journalize the original entry, the adjustment on December 31, 1974, and the adjustment on December 31, 1975, assuming the revenue is recorded:

(a) A portion as earned revenue, the remainder as unearned revenue.
(b) All as earned revenue.
(c) All as unearned revenue.

10-6. The Craft Company issued $1,000,000, 3% bonds at 93 on May 1, 1973. The bonds are dated March 1 and interest is payable on March 1 and September 1. The bonds mature on March 1, 1979. On July 1, 1974, the Craft Company purchased the bonds on the open market at 95 plus interest and retired them.

REQUIRED. Journalize all entries related to the bonds for 1973 and 1974.

10-7. The Ballay Company sold $300,000 of serial bonds on January 1, 1974. Bonds of $30,000 will be retired at the end of each year. Interest of 4% is to be paid semiannually on June 30 and December 31.

REQUIRED

(a) Journalize all entries related to the bonds for 1974 and 1975.

(b) On December 31, 1975, in addition to the bonds, The Ballay Company has the following liabilities:

Accounts payable	$22,000
Salary payable	40,000
Deferred tax liability	3,800

Prepare the liability section of the balance sheet.

Conceptual Problems

10-8. Hartman Hotel Supplies, Inc. must pay a 5% sales tax on all sales it makes. The sales tax is collected from customers but is recorded together with the sales price in the Sales account. When merchandise is returned by the customers, the sales price and sales tax are refunded and recorded together in the Sales Returns and Allowances account. At the end of the first quarter of the year, the Sales account had a balance of $523,493.25 and the balance in the Sales Returns and Allowances account was $588.00.

REQUIRED

(a) Determine the net amount of sales tax owed by Hartman Hotel Supplies, Inc.

(b) Journalize the correcting entry necessary to record the liability for sales tax.

(c) Discuss the nature of the Sales Tax Payable account and why sales taxes should not be included in sales.

10-9. Albert Waddell signed a ten-year lease on January 1, 1973 to rent a small office building. His annual rent payments were to be based on gross sales. On sales up to $100,000, the rate was to be 5%. On any sales in excess of $100,000, the rate was to be 3%. Since Waddell was just starting in business, the lease provided for a minimum annual rental of $4,500 for each of the first five years. Gross sales by years were as follows:

1973—	$ 80,000
1974—	87,000
1975—	95,000
1976—	104,000
1977—	120,000
1978—	105,000
1979—	136,000
1980—	172,000
1981—	150,000
1982—	88,000

REQUIRED. Compute the amount of rent to be paid each year under the terms of the lease. Should the lease be recorded as an asset and liability on the books of Waddell? If so, at what amount, and if not, why?

10-10. The Jordan Manufacturing Company sold at face value $100,000 convertible bonds at 4½% interest. Each $1,000 bond is convertible into 20 shares of common stock. The stock is presently selling at $45 per share. At the time of the sale, the market rate of interest for similar nonconvertible bonds was 6½%.

REQUIRED
(a) Why might Jordan have decided to obtain funds by selling convertible bonds rather than some other means of obtaining funds?
(b) Why would investors be willing to purchase bonds which pay 4½% when they could buy similar bonds paying 6½%?
(c) Does the sale of the 4½% bonds at face constitute a premium or discount? Explain.
(d) How should the bond issue be shown on the balance sheet of Jordan?

10-11. The Stukel Corporation is about to issue bonds on the open market. The financial officers of the company have reviewed the current situation in the bond market and have determined that bonds which are comparable to the ones they plan to issue are presently yielding 7% interest. The officers believe this rate to be higher than they would like to pay. They believe that a rate of 5% is as high as they can offer based upon the projected income to be earned from the funds generated. The company needs a total of $800,000 in order to fund the planned projects.

REQUIRED. Discuss the relative merits of issuing 5% bonds in a market that is yielding 7%.

10-12. The Dry Well Service Station has instituted a promotional program in which a coupon is given to each customer who buys ten or more gallons of gasoline. When the customer has accumulated five coupons they may be redeemed for a free car wash at Phil's Car Palace. The Car Palace has agreed to the promotional set-up and is charging the service station $1.00 per car wash. The service station paid for 2,000 car washes on January 2, 1974. During the month of January 50,000 gallons of gas were pumped, 3,800 coupons were issued and 2,500 were redeemed. During the month of February 65,000 gallons were pumped, 5,200 coupons were issued and a total of 5,400 coupons were redeemed. The service station assumes that all coupons will be redeemed due to the fact that all its customers are from the local area.

REQUIRED
(a) What is the value of one coupon?
(b) Compute the total cost of the promotion for the first two months.
(c) At the end of the second month what liability does the service station have for unredeemed coupons?
(d) Prepare journal entries relative to the promotional program for the first two months.

11

OWNERS' EQUITY

EDUCATIONAL OBJECTIVES

The material in this chapter is designed to achieve several educational objectives. These include:

1. A knowledge of the components of the Owners' Equity section of the balance sheet together with the legal implications of the different components.
2. An insight into the different forms of business organization, i.e., proprietorships, partnerships, and corporations, and the unique accounting procedures required for each type.
3. An awareness that the distribution of net income in the corporate form may be either cash dividends or stock dividends, and that cash dividends are income to the stockholders while stock dividends are not income.
4. A knowledge of the nature of treasury stock, the effect on the Owners' Equity section, and the procedures necessary to adequately account for treasury stock transactions.

In the preceding chapter, all equities were described as the sources of assets. In that chapter, liabilities were presented as one major category, or source. In this chapter, the other major source, that is, owners' equity, will be presented.

The owners' equity section of the balance sheet is important for several reasons:

1. Information is provided to the present owners relative to their proportion of ownership in the business. The Owners' Equity section is often referred to as the *net worth* of the owners in the business.

2. Information is provided for the benefit of prospective investors. They may want to determine their expected share of ownership in the firm.

3. Creditors are interested in this category because they expect the owners to bear a major share of risk in the business. If the owners have only contributed a nominal amount of the total capital relative to the creditors' share, then the risk may in fact be shifted to the creditors.

In the preceding chapter, the criteria for distinguishing debt and equity were presented. At this point, those factors should be reviewed in the context of indicating owners' equity characteristics.

A basic difference between liabilities and owners' equity, but one which cannot be considered a criterion for determining classification, is that owners' equity is a residual. That is, in the equation Assets = Liabilities + Owners' Equity, it is clear that the claims of creditors precede those of the owners. In addition, the claims of creditors are fixed in amount and even though the business may be successful, they will not receive any greater payment for their contribution of resources.

The equity of the owners, on the other hand, may be significantly increased as a result of the successful operation of the business. Therefore, when the business is successful, the owners are in a favorable position relative to the creditors. If the business does not operate successfully, the creditors will still receive repayment of the funds loaned to the business. This is true except in the case of liquidation through bankruptcy. In this event, the claims of creditors may not be fully paid; thus nothing would be available for the owners. This is an example of the residual nature of the owners' equity. Thus the owners may receive substantial reward or they may lose part or all of their investment.

Sources of Owners' Equity

There are two primary sources of owners' equity — contributions by the owners and earnings retained in the business. A secondary source is in the form of donations to the business.

Contributions by the Owners. When a firm is organized, it is common practice for the owners to contribute resources to the business. The contribution is most frequently in the form of cash, but it may be in the form of any of the assets previously discussed, such as inventory, buildings, equipment, patents, etc. If the contribution is in any form other than cash, a fair market value must be determined in order to appropriately record the asset value and the corresponding amount of owners' equity. This is an application of the accounting rule that all assets should be recorded, at the time of acquisition, at their cost or fair market value, whichever is most objectively determinable.

Earnings Retained in the Business. As previously explained, when revenue is earned by the business the equity of the owners is increased, and when expenses are incurred the equity of the owners is decreased. If the net effect is an increase, the business has earned a profit. The profit may be distributed or retained in the business. The statement that profits are retained in the business means that resources, in the form of cash or other assets, are being retained; profits are the source of those resources.

Donations to the Business. A portion of the assets of the business may have been derived from donations. One fairly common donation is in the form of building sites contributed by a city or industrial development commission in order to encourage a business to locate in a particular community. As previously stated, the property is recorded at its fair market value with a corresponding increase in the owners' equity.

FORMS OF BUSINESS ORGANIZATIONS

In this section the accounting activities for a proprietorship, a partnership, and a corporation will be presented.

Proprietorships

A proprietorship is a single-owner form of business enterprise. As such, from a legal point of view, the assets belong to the owner, and all profits are increases in the net worth of the owner. From an accounting point of view, the personal affairs of the owner are assumed to be separate from the business activities. The business constitutes a separate accounting entity even though, legally, this is not the case.

There are three basic types of transactions which affect the owners' equity. These transactions are (a) original contribution, (b) periodic earnings, and (c) distribution to the owners. In the case of a proprietorship, these transactions are recorded as follows.

Original Contribution. Assume that R. Clay organized his own firm, Clay Distributing Company, by depositing $15,000 in a separate bank account. The entry would be:

Cash	15,000	
R. Clay, Capital		15,000

Note that the equity is specified as to the individual owner.

Periodic Earnings. Assume that during the year, net earnings amounted to $12,000 which resulted from revenue of $98,000 and expenses of $86,000. The entry to close the revenue and expense to the capital account would be:

Revenues	98,000	
Expenses		86,000
R. Clay, Capital		12,000

Distributions to the Owner. Assume that during the year, the owner withdrew $800 per month from the business for his personal expenses. The usual procedure is to record these withdrawals in a separate account during the year for informational purposes. The entry each month to record these withdrawals would be:

R. Clay, Drawings	800	
Cash		800

Note that the amount withdrawn for personal use is recorded as a drawing and not an expense. The $800 may be considered a salary by Clay, but it is not deducted as salary expense in calculating net income. The withdrawal is not salary because there is no transaction between two parties. The owner and business are the same; thus the owner could determine net income by adjusting his salary to any level he desired. At the end of the year, the drawing account is closed to the capital account as follows:

R. Clay, Capital	9,600	
R. Clay, Drawings		9,600

The preceding entries reflect the fact that all transactions affecting the proprietor's equity are eventually recorded in the one account, R. Clay, Capital. Therefore, the balance sheet presentation of the proprietorship at the end of the period would be:

R. Clay, Capital	17,400

The ending balance is calculated from the initial investment, $15,000 plus net income, $12,000, minus withdrawals of $9,600.

Partnerships

A partnership is an arrangement whereby two or more individuals agree to operate a business for their mutual benefit. The agreement among the partners is, or should be, written in a document called the *Articles of Partnership*. This document generally contains information relative to:

1. The nature of the business.
2. The amount of investment to be made by each of the partners.
3. A statement as to how net earnings will be distributed among the partners.
4. A statement as to how net losses will be distributed among the partners.
5. The role of the individuals in the management of the business.
6. The conditions and procedures for dissolving the partnership.

In the absence of such an agreement, the law pertaining to partnerships states that profits and losses will be shared equally among the partners, irrespective of the amount of their individual investment and contribution to the daily affairs of the business. Since the conditions governing each partnership may be different, the accounting must be based upon the specific partnership agreement.

The illustration of partnership accounting is based upon the following assumptions:

1. Partners: William Andrews, James Hunter, and Donald Smith.
2. Investment: Andrews, $20,000 cash. Hunter, $10,000 cash and $10,000 (fair market value) equipment. Smith, $20,000 (fair market value) building.
3. Distribution of profits: Each partner to receive 6% return on investment, the balance to be distributed: Andrews, 40%; Hunter, 35%; and Smith, 25%.
4. Withdrawals during the year: Each partner received the following amounts as salary payment: Andrews, $9,600; Hunter, $8,400; and Smith, $7,200.
5. Net income for the year: $35,000.

Original Contribution. The entry for the original investment is:

Cash	30,000	
Equipment	10,000	
Building	20,000	
W. Andrews, Capital		20,000
J. Hunter, Capital		20,000
D. Smith, Capital		20,000

As in the case of the proprietorship, the capital is specified by individual accounts.

Periodic Earnings. In the distribution of net income, each partner is allocated 6% as interest on his investment. The remaining $31,400 is allocated according to the agreement:

		Profit and loss ratio		
		40%	35%	25%
		Andrews	Hunter	Smith
Net income	$35,000			
— Interest on original investment				
(60,000 × 6%)	3,600	$ 1,200	$ 1,200	$1,200
Remaining profit distributed				
in profit and loss ratio	$31,400	12,560	10,990	7,850
Net earnings distributed		$13,760	$12,190	$9,050

If the revenues for the year were $250,000 and the expenses $215,000, the entry to close these accounts to the partners' capital accounts would be:

Revenues	250,000	
Expenses		215,000
W. Andrews, Capital		13,760
J. Hunter, Capital		12,190
D. Smith, Capital		9,050

Distributions to the Owners. The journal entry to close the Drawings accounts for the salary payments would be:

W. Andrews, Capital	9,600	
J. Hunter, Capital	8,400	
D. Smith, Capital	7,200	
W. Andrews, Drawings		9,600
J. Hunter, Drawings		8,400
D. Smith, Drawings		7,200

Balance Sheet Presentation. The capital section of the balance sheet would be:

W. Andrews, Capital	24,160
J. Hunter, Capital	23,790
D. Smith, Capital	21,850
	69,800

The amounts on the balance sheet were derived from the following Statement of Partners' Capital:

	Andrews	Hunter	Smith	Total
Capital, 1/1/74	$20,000	$20,000	$20,000	$60,000
Add: Net earnings	13,760	12,190	9,050	35,000
	$33,760	$32,190	$29,050	$95,000
Less: Withdrawals	9,600	8,400	7,200	25,200
Capital, 12/31/74	$24,160	$23,790	$21,850	$69,800

Corporations

The accounting for a proprietorship and a partnership have several common characteristics inasmuch as those two forms of business are similar, except that in the proprietorship there is one owner and in the partnership there are several owners. There is, however, a notable difference between those two forms of business enterprise and a corporation.

One of the basic differences between corporations and other forms of business is that, generally, the number of owners is substantially greater than in a partnership. This number may run into the hundreds or thousands. There is also a basic legal difference in that the owners of a corporation normally have no liability beyond their investment in the business, whereas a proprietor and a partner have a liability to the full extent of their personal wealth. Because of the limited liability factor, there are certain legal restrictions on corporations that do not apply to proprietorships and partnerships.

The limited liability of owners is justified because, legally, a corporation is a separate entity from those who contribute the capital. As an entity, the corporation has many of the same rights as individuals. Among these are: the right to engage in business in its own name; the right to own property; and the right to legal action in its own name.

The legal status of a corporation, combined with the large numbers of owners and its perpetual existence, often makes it easier to raise the large amount of capital which is necessary in many modern businesses. Investors, both creditors and owners, are generally more inclined to provide funds to an organization which will not be terminated upon the death, or withdrawal from the firm, of one individual, as in the case of proprietorships and partnerships.

One important aspect of a corporation is that the state must grant a charter. The charter specifies the detailed conditions under which the corporation may operate, such as what business activities are permitted, types and amount of capital stock to be issued, and the method of electing officers. The number of shares a company is permitted to sell is called the *authorized stock*. The number of shares actually sold is called the *issued stock*.

In a corporation, ownership is evidenced by a document referred to as a *stock certificate*, and the owners are referred to as *stockholders*. Some certificates have a value determined by the board of directors, in which case the stock has a *par value*. Certificates may be issued without a face or par value and these are called *no-par stock*; however, the laws of some states prohibit the issuance of no-par stocks. In these states the board of directors must specify a particular value called *stated value*. Thus par value and stated value stocks are basically the same for accounting purposes.

Types of Stock. All corporations issue shares of stock referred to as *common stock* and, in addition, many issue shares known as *preferred stock*.

Common stock. The holders of common stock have certain legal rights. Among the rights are:

1. The right to share in any *distribution* of profit made to common stockholders on a per share basis. However, the stockholder does not have any right to the earnings until the board of directors officially declares a dividend. This is a major point of difference between ownership rights in a proprietorship or partnership and a corporation.

2. The right to elect the board of directors and to vote on certain other major decisions. However, the right may be meaningless to an individual, because the proportion of ownership may be insignificant relative to the total outstanding shares. While all stockholders are invited to the annual meeting of their company, most do not attend and the present management requests the right to vote by *proxy* the shares of those who will not be present.

3. The right to maintain their present percentage of ownership when the company issues additional shares of stock. This right is known as the *pre-emptive right* and means that the present stockholders must be given the opportunity to acquire, on a *prorata* basis, the new shares.

4. The right to share in any assets available for distribution if the corporation is liquidated. The stockholders receive payment only after the claims of all other equity holders have been fully paid.

The basic rights outlined here may be limited, or varied, by the terms of a particular stock certificate or applicable state laws. In most cases, preferred stockholders do not have all of these rights, but they have some priority as to other rights.

Preferred stock. The term *preferred stock* means that holders of such stock certificates have certain rights which have priority over holders of common stock. Generally, these preferences pertain to dividends and liquidation. Preferred stockholders, even though considered to be owners, normally must be paid a dividend before the common stockholders are eligible to receive any payment. However, as in the case of common stockholders, they do not have any legal right to dividends prior to their declaration by the board of directors.

The amount of the preferred dividend is generally a fixed percentage of the par or stated value of the shares. In other cases, the dividend is specified as a dollar amount per share. Sometimes the dividend is *cumulative*, which means that if the specified dividend is not declared and paid in a given year, the amount is accumulated. All accumulations must be paid to the preferred stockholders before payment can be made to the common stockholders.

Preferred stock may also be *participating*. This means that, once a specified amount of dividends has been paid to common stockholders, the preferred stockholders participate, in a specified ratio, with the common stockholders in any additional distribution of dividends. As in the case of bonds, preferred stock may be *callable*, at the option of the company. Preferred stock may also be *convertible* into shares of common stock.

Original Contribution. Assume that the Howard Company was legally organized with authorization to sell 1,000 shares of $100 par value common stock and 500 shares of $50 par value, 7%, preferred stock. On May 1, 1974, Howard Company received cash of $20,000 and equipment with a fair market value of $15,000 in exchange for 300 shares of common and 100 shares of preferred stock. The journal entry to record this transaction is:

Cash	20,000	
Equipment	15,000	
Common stock		30,000
Preferred Stock		5,000

The entry is more complex if the value of the assets received is different from the par value or stated value of the stock. Since the market price of stock is continually changing, there is normally a difference between the market price and the par value. Therefore, assume that in the previous example, the cash received was $21,500 because the market price of the common stock was $105 per share. The journal entry would be:

Cash	21,500	
Equipment	15,000	
Common Stock		30,000
Preferred Stock		5,000
Capital in Excess of Par		
Value — Common Stock		1,500

Whenever shares of stock are issued for more or less than the par or stated value, the premium or discount is recorded in an account separate from the stock. The premium or discount is related to the type of stock involved and is shown on the balance sheet as a part of owners' equity. (See balance sheet presentation at end of chapter.) In most states, stock will not be sold for less than par or stated value. In the event that the firm becomes insolvent, the stockholders would be legally liable to the creditors for the amount of discount.

Periodic Earnings. In a corporation, the Owners' Equity section of the balance sheet reflects the various types of equity. Therefore, unlike the previous illustrations of a proprietorship and partnership, the periodic earnings of the firm are reflected in a separate account from the original contribution of the owners. The account used for this purpose is *Retained Earnings*.

In the closing process, the revenue and expense accounts are summarized and the net difference is transferred to Retained Earnings. For example, if the revenues for the period were $300,000 and the expenses were $298,000, the entry to record the transfer to Retained Earnings would be:

Revenues	300,000	
Expenses		298,000
Retained Earnings		2,000

The closing process for a corporation may appear to generate less information than in the case of a partnership where the amount allocated to each individual is determined. Actually, the same basic information is provided in corporate reports; however, the format is different. In a corporation this information is in the form of earnings per share of common stock. Each stockholder may multiply the earnings per share times the number of shares owned to determine his total portion of net income. The stockholder has no right to withdraw his portion of income; therefore, the calculation is essentially for informational purposes. In addition, the market price of the stock is usually a function of the earnings per share of the company.

In Chapter 13 the details involved in calculating earnings per share will be discussed. The purpose at this point is to illustrate briefly the nature of the calculation. If the previous assumptions relative to net income and stock issued are continued, the calculation of earnings per share would be:

Net income	2,000
Less: Dividends due preferred stockholders ($50 × 7% × 100 shares)	350
Net income for common stockholders	1,650

$$\text{Earnings per share} = \frac{\$1,650}{300 \text{ shares}} = \$5.50 \text{ per share}$$

The Accounting Principles Board in Opinion Number 15 stated that earnings per share must be shown on the income statement of a corporation. This requirement indicates the importance attached to this calculation as a source of information for the users of financial statements.

Distributions to the Owners. In a corporation, distribution of earnings is referred to as *dividends*. This distribution to the owners may involve:

1. Distribution of corporate assets such as cash, property, etc.
2. Issuance of additional shares of stock, i.e., stock dividend.

The payment of dividends is restricted, by the laws of most states, to the amount of retained earnings. The laws are intended to provide protection for the creditors. This protection is important because of the limited liability of the stockholders. Without these restrictions, the owners could declare and pay themselves dividends such that their investment in the firm would be eliminated. If this were permitted to occur, the owners would eliminate the possibility of a loss since their investment in the firm would be zero.

Cash dividends. The payment of a cash dividend is the most common method of distributing a portion of the earnings to the stockholders. In the earlier example, the earnings were determined to be $5.50 per share of common stock. The amount of dividends, if any, declared by the board of directors, may have little relationship to the earnings per share.

As has been stated before, the Retained Earnings account is merely a record of earnings retained in the business since its legal organization.

The actual earnings resulted in an increase in assets; however, the assets are not necessarily in the form of cash. The cash collected from sales may have been used to purchase more inventory, equipment, buildings, or any other asset. Therefore, cash may not be available to pay dividends even though the firm has a substantial balance in the Retained Earnings account.

The board of directors may declare a dividend less than $5.50 for reasons other than insufficient cash, e.g., the cash is needed to retire outstanding bonds. The board is given the power, and responsibility, to make decisions relative to dividends. This responsibility includes compliance with legal requirements as well as meeting the needs of the business.

In order to illustrate the accounting treatment of a cash dividend, assume that the board of directors declared, in view of legal and firm requirements, a cash dividend of $2.00 per share on the 300 shares of common stock. The entry to record the declaration is:

Retained Earnings	600	
Dividends Payable		600

Note that the declaration results in a decrease of retained earnings and the creation of a current liability, because once the dividends are declared, they must be paid. In addition, there is a necessary processing time between the date dividends are declared and the date they are paid. The time lag is necessary in order for the corporation to bring its stockholder records up-to-date and prepare the checks. The date on which the board declares the dividend is called the *declaration date*. The cut-off date for inclusion on the stockholder list is called the *date of record*. This date is usually two or three weeks after the date of declaration. At the time of payment, the entry is:

Dividends Payable	600	
Cash		600

Stock dividends. The board of directors may declare a dividend when there is no cash available for payment. The declared dividend may be in the form of additional shares of common stock to be issued to the stockholders, without any payment from them in return. Such a dividend is referred to as a *stock dividend*.

A stock dividend permits the firm to give the shareholders some evidence of their proportionate interest in the retained earnings. Since there is no need to distribute cash or other property, the assets of the firm remain the same. Thus a stock dividend may be issued primarily for the appearance of giving the stockholders something they did not have before. What they actually receive are more pieces of paper representing their same respective interests in the firm. In order to obtain cash, they must sell the new shares thereby decreasing their relative ownership interest.

The question of whether stock dividends are income to the stockholders has been legally resolved. In the historic case, Eisner versus Macomber, 252 U.S. 189, in which the Internal Revenue Service argued for the taxation of stock dividends as income, Mr. Justice Pitney asserted that:

> A stock dividend really takes nothing from the property of the corporation and adds nothing to the interests of the stockholders. Its property is not diminished and their interests are not increased . . . the proportional interest of each shareholder remains the same. The only change is in the evidence which represents that interest, the new shares and the original shares together representing the same proportional interests that the original shares represented before the issue of the new ones.

As a result of that opinion, stock dividends are not legally income and, therefore, not taxable. A stock dividend, however, does require recognition in the owners' equity accounts in order for the financial statements to convey the appropriate information.

When a stock dividend is issued, retained earnings must be transferred to the permanent capital accounts. Otherwise, a stock dividend could be issued and, later, a cash dividend declared based upon the same retained earnings. Many stockholders view a stock dividend as a distribution of earnings in the amount of the fair market value of the shares received. Therefore, the amount transferred from retained earnings should be equal to the fair market value of those shares.

The following illustration assumes that a 4% stock dividend was issued on the 300 shares of $100 par value stock previously described, at a time when the market price of the stock was $110. The calculation of the number of new shares is $300 \times .04 = 12$ shares. The journal entry would be:

Retained Earnings (12 × 110)	1,320	
Common Stock (12 × 100)		1,200
Capital in Excess of Par Value — Stock Dividend (12 × 10)		120

When the shares of stock distributed are less than 20–25% of the number previously outstanding, the market price will probably remain near the level which existed prior to the dividend. The 20–25% level is the guideline established by the Accounting Principles Board. This is another reason for transferring the fair market value from retained earnings. If the distribution is greater than 20–25%, the market price will probably decrease and the distribution is more appropriately labeled a stock split instead of a stock dividend.

Stock split. A stock split is similar to a stock dividend in that additional shares of stock are issued, without cost, to the stockholders; however, the purpose is to reduce the market price per share in order to increase the salability of the shares. The intent of a stock split, therefore, precludes the need to record a transfer from Retained Earnings to the permanent capital accounts. The only accouting action required is a notation concerning the new number of shares outstanding.

Appropriated Retained Earnings. As discussed previously, the board of directors may believe that dividends should be limited because of anticipated needs of the firm or legal restrictions. In order to convey this information to the stockholders, the board of directors may designate a portion of the retained earnings as being appropriated for these purposes. Actually the board has the authority and responsibility to limit dividends without any formal appropriation. The appropriation has no effect on the assets of the firm; however, because of the appropriation, the stockholders may be more receptive to a smaller dividend.

To illustrate the accounting necessary as a result of formal action by the board of directors, assume that of total retained earnings of $50,000, the following amounts have been designated for special purposes:

Retirement of bond issue	$25,000
Purchase of new equipment	5,000
Miscellaneous contingent needs	5,000
	$35,000

The journal entry to record the appropriation is:

Retained Earnings	35,000	
Retained Earnings Appropriated		
for the Retirement of Bonds		25,000
Retained Earnings Appropriated		
for the Purchase of Equipment		5,000
Retained Earnings Appropriated		
for Miscellaneous Contingencies		5,000

When the special need no longer exists, the appropriated amount is returned to unappropriated retained earnings. For example, assume that the bonds are retired. The entries would be:

Bonds Payable	25,000	
Cash		25,000
Retained Earnings Appropriated		
for the Retirement of Bonds	25,000	
Retained Earnings		25,000

Note that the retirement of the bonds required the use of cash. The appropriated portion of retained earnings did not have any effect on the assets and liabilities. Thus it can be seen that the total retained earnings is the same after the bonds are retired as before the appropriation was made.

Treasury Stock. There are circumstances in which a firm may purchase shares of its own stock. If the shares thus acquired are not retired they are referred to as *treasury stock*. The purchase of treasury stock gives rise to some unique accounting problems.

The acquisition of its own stock does not constitute an investment by the company. Thus treasury stock is not shown as an asset on the balance sheet. Rather, the transaction may be viewed as a temporary liquidation of a portion of the outstanding stock and is generally shown on the balance sheet as a reduction of total stockholders' equity.

As an illustration of the recording process, assume that XYZ Company purchased 100 shares of its own stock in the open market at $25. A common method of recording treasury stock is at the cost to the firm. Thus the entry to record the purchase by XYZ Company of its stock is:

Treasury Stock	2,500	
Cash		2,500

If at a later date the stock is sold for $30, the entry would be:

Cash	3,000	
Treasury Stock		2,500
Capital in Excess of Cost — Treasury Stock		500

The reason for the credit to Capital in Excess of Cost—Treasury Stock is that the excess over cost is considered an additional investment in the firm. In addition, both from accounting and legal points of view, a company cannot earn a profit through the purchase or sale of its own stock. When the balance sheet is prepared, the amount of issued common stock includes the par value of the treasury stock. The cost of the treasury stock is a reduction of total stockholders' equity. Since the transaction has the effect of reducing the owners' permanent equity, there may be certain legal restrictions on the amount of treasury stock a firm may acquire.

Another method of accounting for treasury stock transactions is called the *par value method*. In this method the treasury stock is recorded at the par value, and the difference between the par value and the amount paid is a reduction of the capital in excess of par value attributable to the treasury shares and any remaining difference is a reduction of retained earnings.

For example, if XYZ Company stock had a par value of $15 and the stock was originally sold for $21, the acquisition of 100 shares at $25 would be recorded as:

Treasury Stock	1,500	
Capital in Excess of Par Value	600	
Retained Earnings	400	
Cash		2,500

When the stock was sold for $30, the entry would have been:

Cash	3,000	
Treasury Stock		1,500
Capital in Excess of Par Value		1,500

If the par value method is used, the treasury stock is normally shown as a reduction of the class of stock rather than a reduction of total stockholders' equity.

Balance Sheet Presentation

The purpose of the following balance sheet is to illustrate the Stockholders' Equity section of a corporate balance sheet. Therefore, the amounts were not derived from the transactions used in the previous examples:

Stockholders' Equity

Preferred stock — 5% cumulative, $100 par value (authorized 40,000 shares, issued 20,000)		$ 2,000,000
Common stock — $10 par value (authorized 1 million shares, issued 600,000 of which 5,000 are held as treasury stock)		6,000,000
Capital in excess of par value — common stock		98,000
Retained earnings:		
Unappropriated	$1,600,000	
Appropriated for bond retirement	500,000	
Appropriated for building expansion	100,000	
Total retained earnings		2,200,000
Less: Cost of treasury stock (5,000 shares)		(100,000)
Total Stockholders' Equity		$10,198,000

SUMMARY

The purpose of the Owners' Equity section of the balance sheet is to provide information relative to the various ways the owners have provided funds for the business, i.e., purchase of preferred stock, common stock, payments in excess of par or stated value, and profits retained in the business. This information is useful to the owners, prospective owners, and creditors.

The forms of business organizations include proprietorships, partnerships, and corporations. Each of these requires certain unique accounting treatments relative to owners equity; however, since the firm is viewed as a separate entity, most accounting practices are the same. Each business enterprise has transactions relative to the contribution of capital, earnings of the business, and distributions of those earnings.

QUESTIONS

11-1. What is the importance of the Owners' Equity section of the balance sheet?

11-2. What are the primary sources of owners' equity? Do they always flow into the business in the form of cash?

11-3. What is meant by the proprietorship form of business? What accounts compose the Owners' Equity section of a proprietorship form of business?

11-4. Assume Bill Smith invested $25,000 in organizing his own business. During the year revenues totaled $18,500, and expenses amounted to $12,800. The owner also withdrew $500 per month for personal expenses during the year. Determine the ending balance in Smith's Capital account.

11-5. What is meant by the partnership form of business organization? Is a written agreement required to begin a partnership?

11-6. List the types of information included in the *Articles of Partnership*. In the absence of such an agreement, how are profits and losses distributed?

11-7. Joan, Sharon, and Dolly formed a partnership with each investing assets having a fair market value of $30,000. The *Articles of Partnership* state that profits will be divided as follows: each partner to receive a 5% return on her investment and the balance distributed 30% to Joan, 35% to Sharon, and 35% to Dolly. If net income during the year totaled $45,000, how would it be divided among the three partners?

11-8. Define the term *limited liability* as it applies to owners of a corporation.

11-9. What rights does a corporation have in common with an individual?

11-10. Distinguish between par value and no-par value stock.

11-11. Two types of stock are issued by corporations—common and preferred. List the rights inherent in owning each class of stock.

11-12. Describe *cumulative* and *participating* preferred stock.

11-13. Bibb Corporation exchanged 200 shares of its $40 par value common stock and 50 shares of its 6%, $100 par value preferred stock for a tract of land and a building. The land had a fair market value of $4,000 and the building had a fair market value of $9,000. Prepare the journal entry for the transaction.

11-14. Assuming the land described in Question 13 has a fair market value of $6,000, and the market price of the common stock is $50 per share, prepare the journal entry.

11-15. What is the purpose of the Retained Earnings account in the Equity section of a corporation's balance sheet?

11-16. Why is the payment of dividends in most states restricted to the amount of retained earnings?

11-17. Slade Corporation has been in business for five years. During this period, they have realized a net income in each year of operations. At the end of the five-year period, Retained Earnings has a credit balance of $38,500. What observations can you make concerning the cash balance of Slade Corporation?

11-18. How are the assets of a corporation affected by a cash dividend and a stock dividend.

11-19. Distinguish between a stock dividend and a stock split.

11-20. What is the objective of an appropriation of retained earnings?

PROBLEMS

Procedural Problems

11-1. The Zigmond Company finished its first year of operations on December 31, 1974. For the year, the company had revenues of $78,000 and expenses of $57,000.

REQUIRED. Prepare the necessary closing entries under the following assumptions:
(a) The business was organized as a proprietorship by J. Zigmond who withdrew $100 per month from the business during the first year of operations.
(b) The business was organized as a partnership with three partners; J. Zigmond, S. Scott, and T. Stacy. The partners share profits in the following ratio: Zigmond 40%; Scott 35%; Stacy 25%. Each partner withdrew $500 during the year.
(c) The business was organized as a corporation. Dividends of $8,000 were paid during the year and recorded in a Dividend account.

11-2. The Creekside Corporation was organized on January 1, 1974. Its charter authorized it to issue 5,000 shares of $10 par value, 5% preferred stock and 10,000 shares of $15 par value common stock.

REQUIRED
(a) Prepare the journal entry Creekside Corporation would make for the sale of 4,000 shares of preferred stock and 8,000 shares of common stock. They received $60,000 for the preferred and $136,000 for the common.
(b) Prepare the entry for the sale of the remainder of the shares, both common and preferred, to George Stone for a building valued at $40,000.

(c) For the first three years of operations, the corporation earned a total of $48,000. During the first two years the company did not pay dividends. The board of directors decided to pay a dividend as of the end of the third year of operations. The amount of the dividend declared was $13,500. If the preferred stock is cumulative and nonparticipating, how much would the preferred and common shareholders receive?

(d) Prepare the Equity section of the Creekside Corporation balance sheet as of January 1, 1977, the beginning of the fourth year of operation.

11-3. Red, White, and Blue are partners. Their first act in forming the partnership was to retain a lawyer in order that the *Articles of Partnership* could be formulated. In terms of sharing profits and losses, the agreement indicated the following plan:
- Each partner is to receive 6% interest on his capital balance as of the beginning of each year.
- The remaining profits are then divided in the following manner: Red, 20%; White, 50%; Blue, 30%.

The partners also agreed that each would withdraw a certain amount of cash from the business each month. The amount of drawings for Red, White, and Blue are $500, $700, and $800 per month, respectively. In order to begin the business, Red invested $40,000 in cash, White invested a building with a market value of $80,000, and Blue invested machinery and equipment valued at $50,000. During the first year of operations, net income amounted to $25,000 and in the second year, net income was $45,000.

REQUIRED. Prepare a statement of partners capital at the end of the first and second year of operations.

11-4. The Midwestern Corporation received a charter early in 1974. The charter authorized the corporation to issue 20,000 shares of $5.00 par value common stock and 5,000 shares of $10 par value, 5% preferred stock. The preferred stock is neither cumulative nor participating. The following selected transactions involve the owners' equity of Midwestern Corporation:

(a) Issued 10,000 shares of common stock for $8 per share, and 3,000 shares of preferred stock for $12 per share.

(b) Issued 500 shares of common stock to C. Darrow for legal services rendered in connection with organizing the corporation. Mr. Darrow usually receives $400 per day for his services and he spent ten days working for Midwestern Corporation.

(c) Issued the remaining shares of preferred stock for machinery and equipment valued at $23,500 by the board of directors.

(d) Declared an annual dividend on preferred and a $0.50 per share dividend on common stock.

(e) Paid the dividends declared in item (d).

(f) Established a reserve for plant expansion in the amount of $15,000.

(g) Issued a 10% stock dividend on the common stock. The current market price of the common stock was $10 per share.

REQUIRED. Prepare the journal entries for the preceding transactions assuming the corporation has operated profitably throughout its existence.

11-5. The Wilson Metal Corporation is authorized to issue 10,000 shares of $50 par value common stock and 5,000 shares of $100 par value, 6%, cumulative, nonparticipating preferred stock. The transactions listed as follows are those affecting the owners' equity accounts for the first three years of operations:

1974

Jan. 3   Sold 7,000 shares of common stock for $400,000 cash and issued 4,000 shares of preferred for land valued at $200,000 and a building valued at $300,000.

Dec. 31   Income for the first year of operations amounted to $50,000. Revenues totaled $125,000 and expenses amounted to $75,000. No dividends were declared or paid.

1975

Mar. 15   Issued 1,000 shares of common stock for $55,000 cash.

June 30   The corporation issued bonds and appropriated retained earnings for the retirement of bonds in the amount of $40,000.

Dec. 31   Income for the year amounted to $105,000. Revenues totaled $270,000 and expenses totaled $165,000. No dividends were declared or paid.

1976

July 1   Redeemed the bonds issued in 1975.

Nov. 15   Appropriated retained earnings for plant expansion in the amount of $60,000.

Dec. 31   Income for the year amounted to $140,000. Revenues totaled $320,000 and expenses totaled $180,000. Dividends were declared and paid to preferred shareholders for the year plus dividends in arrears. The common shareholders received a $4 per share dividend. (Assume the dividends were declared and paid on the same day.)

REQUIRED. Journalize the foregoing transactions which affect the owners' equity accounts and prepare the Equity section of the balance sheet for Wilson Metal Corporation as of December 31, 1976.

11-6. On January 1, 1974, the Hazard Company had a major fire and all of the company's accounting records were destroyed. Since the company must pay income taxes for the year 1973, the board of directors of Hazard Company comes to you for assistance in arriving at a reasonable estimate of the company's 1973 income. Though the company has no documents to substantiate the following, the bookkeeper does remember a few balances and transactions which took place during 1973:

REQUIRED. Using the pertinent data, prepare a statement estimating Hazard Company's 1973 net income:

Unappropriated retained earnings balance, January 1, 1973	$65,900
Unappropriated retained earnings balance, December 31, 1973	40,900
Established a reserve for plant expansion in June 1973	26,000
Cash in bank December 31, 1973	7,950
Stock dividend declared and issued during 1973	4,000
Issued 100 shares of $100 par value common stock	10,300
Declared and paid a cash dividend during 1973	5,000
Accounts payable, December 31, 1973	8,000

11-7. Rice Corporation has been in business for ten years. During that period, the company has earned substantial income; however, a dividend has never been paid. The reason dividends have not been paid is that the board of directors believes that income from the first ten years of operations should be used to expand the business. Recently the price of Rice Corporation common stock declined. Some investment analysts believe the decline is due to the dividend policy adhered to by the directors. In an attempt to deal with the problem, the directors decided to issue stock as an indication of the corporation's profitable operations. The owners' equity on January 1, 1974 is as follows:

Common stock—$10 par value, authorized 800,000 shares, issued and outstanding 500,000	$5,000,000
Capital in excess of par value	500,000
Retained earnings	1,620,000
Total stockholders' equity	$7,120,000

On January 15, 1974, the directors declared a 10% stock dividend to be distributed on February 15, 1974. The price of the stock was $15 per share on January 15. Subsequent to the declaration of the dividend, the price of the stock began to climb. Because of the success of the first dividend, the directors declared a 5% stock dividend on July 1, when the stock was selling for $25 per share. This dividend was distributed on August 1, 1974. On November 1, 1974 the Rice Corporation stock was split two-for-one, and the authorized number of shares was increased to 1,600,000 shares and the par value reduced to $5 per share. The net income for 1974 amounted to $382,500.

REQUIRED
(a) Prepare the journal entries for the stock transactions indicated.
(b) Prepare the Equity section for Rice Corporation as of December 31, 1974, the company's year end.

Conceptual Problems

11-8. The president of Stable Corporation, which recently began operations, is interested in knowing the effect of various transactions on selected parts of the Owners' Equity section. He asks you to indicate the effect of each event listed on Stable Corporation's:
- Total contributed capital
- Total retained earnings
- Total stockholders' equity

The effect on the preceding items should be stated in terms of increase, decrease, or no effect.
(a) Creation of a reserve for plant expansion.
(b) The sale of $10 par value common stock for $15 per share.
(c) Declaration of a cash dividend on common stock.
(d) A two-for-one stock split.
(e) Payment of a previously declared cash dividend.
(f) Declaration and issuance of a stock dividend on common stock.
(g) Donation of a plant site by a city to Stable Corporation.
(h) A net loss for the year from operations.

11-9. Sly Corporation earned income of $14,000 during the year; however, after closing the revenue and expense accounts, the retained earnings had a debit balance of $52,000. The board of directors, believing that the corporation would soon show a credit balance in retained earnings, declared a $10,000 cash dividend on common stock. A holder of Sly Corporation's cumulative preferred stock believes this dividend is improper but has no basis for his belief. He has come to you for your opinion.

REQUIRED. Discuss the appropriateness of the dividend.

11-10. Harry Lotta and Bill Speed organized the Lotta Speed Corporation. The two men were experts in the computer area and decided to design and manufacture a new model. They decided that the corporate form of organization afforded them the best opportunity for raising the necessary funds to operate the business; however, when it came time to decide upon the type of stock to issue, the men disagreed. Mr. Lotta stated a preference for issuing a large amount of cumulative, nonparticipating preferred stock, and a small amount of common. He reasoned that this mix would allow the two men to easily retain control of the company, and even though the preferred was cumulative, a dividend every year was not required. Mr. Speed favored common stock over preferred because he believed the speculative nature of this product would be much more attractive to the common stock investor. He also stated that they could issue as much common stock to themselves as they wished at a low price and then sell to outsiders at higher prices. The par value on the authorized stock is $50 for common and $100 for preferred.

REQUIRED. Discuss the position held by each individual indicating the strengths and weaknesses in the arguments. What issues do you believe to be most important in deciding upon the best stock mix?

11-11. The Eastinghouse Corporation has accumulated a credit balance in its Retained Earnings account totaling $87,000 during the first five years of operation. Gregory James, president of the company, instructs the chief accountant to appropriate retained earnings of $20,000 per year for five years and to title the account, Reserve for Plant Expansion. At the end of the five year period, the reserve account has a balance of $100,000 and the president orders expansion plans to begin. When the expansion is completed, the chief accountant informs the president that it will be necessary to secure a loan in order to pay the contractors. The president is shocked by this information and informs the accountant to draft a complete report indicating why the $100,000 is not available, especially since the total of both appropriated and unappropriated retained earnings far exceeds that amount. Also include in the explanation the purpose for appropriations of retained earnings.

11-12. The board of directors of Ramit Corporation has worked diligently to make the corporation a profitable entity. Members of the board own approximately 65% of the outstanding common stock and thus make most of the major decisions involving corporate policy. Recently, the directors decided to declare a $2.00 per share dividend on the 65% of the common stock which they own. They also voted to issue 500 shares of common stock to each board member as a reward for his efforts in directing corporate affairs. The directors felt justified in this action since their efforts far exceeded those of other shareholders. Also, they reasoned that such distributions were legal in that a formal vote of all shareholders would produce a 65% to 35% vote in favor of the measures.

REQUIRED. If you were a holder of a portion of the common stock not held by the directors would you have any recourse to the actions outlined? Why or why not?

12

STATEMENT OF CHANGES
IN FINANCIAL POSITION

EDUCATIONAL OBJECTIVES

The material in this chapter is designed to achieve several educational objectives. These include:

1. An awareness of the objective of the statement of changes in financial position and the importance of this statement in light of Accounting Principles Board Opinion Number 19.
2. An understanding of the several concepts of funds and how the choice of the concept influences the preparation of the statement.
3. A knowledge of how the statement of changes in financial position supplements the balance sheet and income statement.
4. A knowledge of the techniques and procedures employed in the analysis of financial transactions before a statement can be prepared.
5. An understanding of the relationship between a cash flow statement and a cash forecast.

As noted in Chapter 2, financial statements are interrelated. The discussion at that point pertained to the balance sheet, income statement, and statement of retained earnings. Each of these statements is intended to convey certain information about the firm; however, there is some important financial information concerning a firm that is not clearly revealed in any of these statements. For example, the acquisition of an asset through the issuance of the firm's stock may not be apparent in the balance sheet. In some cases, this information may be determined by comparing the balance sheets at the beginning and end of the period and noting any changes in the specific asset and stock accounts. Such direct evaluations can only be made, however, when there are no other transactions involving these accounts. When there are several transactions, the sources of the change in the account balance can be determined only through access to the detailed records.

To encourage the presentation of information about the changes in financial position, the Accounting Principles Board issued Opinion Number 3, *The Statement of Source and Application of Funds.* The reason for a statement of source and application of funds (now the statement of changes in financial position) is that the balance sheet and income statement do not provide sufficient information for the user. The *balance sheet* is a statement of the assets, liabilities, and stockholder's equity at a particular date. The *income statement* is the inflow of assets resulting from sales or other income producing activities minus the outflow of assets used to produce the revenue. The *statement of changes in financial position* provides, in a capsule form, information on how the activities of the business have been financed and how the financial resources have been used during the period covered by the statement. Consequently, the funds statement provides information which cannot be easily obtained from either the income statement or balance sheet. Questions such as: Where did the profits go? or, Why were the dividends not larger? may be answered by this statement.

Opinions

In Opinion Number 3, the Accounting Principles Board recommended that a statement of source and application of funds be presented as supplementary information in financial reports. The inclusion of this information was not mandatory, and it was optional whether it should be covered in the audit report of the independent accountant; however, in Opinion Number 19, the Board concluded that information concerning the financing and investing activities of a business enterprise and the changes in its financial position for a period were essential for financial statement users, particularly owners and creditors, in making economic decisions. The Board stated that "when financial statements purporting to present both financial position and results of operations are issued, a statement summarizing changes in financial position should also be presented as a basic financial statement for each period for which an income statement is presented."

Clearly, because of Opinion Number 19, the statement of changes in financial position has been raised in importance and is now a basic statement together with the balance sheet and income statement. The statement is not intended to be a replacement of these other statements but, rather, is intended to serve its own unique function.

Concept

There are three basic concepts of *funds* which have influenced the preparation of the funds statement. These concepts are: cash, working capital, and all financial resources.

Cash. When the concept of funds is cash, the statement becomes a *statement of cash receipts and disbursements*. The purpose of the statement is to explain the change in the cash position of the firm during the period by reflecting the flow of cash into and out of the business. The term *cash flow* means different things depending upon the user, but the most common meaning in accounting is net income *plus* those expense items which did not require any current period cash outlay, e.g., depreciation, depletion, and amortization *less* those revenue items which did not provide cash in the current period, e.g., credit sales and accrued interest. Therefore, a funds statement prepared on a cash basis is broader than the common meaning of cash flow, since the funds statement includes all cash flowing into and out of the business.

Working Capital. The second concept of funds, and the one most commonly followed in the past, is working capital, i.e., current assets less current liabilities. Under this concept, the purpose of the funds statement is to explain the sources and uses of working capital and the resultant change in working capital during the period. As such, the statement would include only those noncurrent items which had an effect on the current accounts, for example, the sale of stock for cash.

All Financial Resources. The use of either the cash or working capital concept limits the usefulness of the funds statement because certain financing and investing activities are excluded. In Opinion Number 19, the Accounting Principles Board concluded that the statement summarizing changes in financial position should be based on a broad concept embracing all changes in financial position. The term used to express this concept is *all financial resources*.

The use of the all financial resources concept eliminates the omission problem of the cash and working capital concepts. All important aspects of the firm's financing and investing activities, regardless of whether cash or other elements of working capital are directly affected, are disclosed. For example, the acquisition of property through the issuance of securities or the conversion of long-term debt to stock would not be disclosed using the working capital concept; however, these transactions are disclosed under the all financial resources concept.

Format. The form of the statement was not specified by the Board in Opinion Number 19. The Board stated that it recognized the need for flexibility in form, content, and terminology. Each entity should adopt the presentation that is most informative in its circumstances. Basically, the format of the statement on all financial resources basis, in which funds are defined as working capital, should include information about:

1. Sources of funds which increase working capital
2. + Sources of financial resources which did not increase working capital

3. = Total funds available
4. − Outflows of working capital
5. − Outflows of financial resources which did not decrease working capital

6. = Change in working capital

Items 2 and 5 are related because the funds generated in item 2 are applied directly to item 5. Thus, items 2 and 5 are equal. Therefore, the change in working capital (item 6) is the difference between items 1 and 4.

A comprehensive example will be presented to illustrate the guidelines and details of the recommended statement of changes in financial position.

XYZ Company
Statement of Changes in Financial Position — Working Capital Basis
for the Year Ended December 31, 1974

Financial resources generated:	
Working capital generated:	
Net income	**$26,000**
Add (deduct) items not requiring or generating working capital in the current period:	
Depreciation expense	**9,500**
Amortization of bond discount	**100**

Amortization of patent	400	
Gain on sale of land	(1,000)	
Gain on sale of machinery	(5,000)	
Working capital generated by operations		$ 30,000
Other sources of working capital:		
Bonds payable sold	$ 5,000	
Common stock sold	5,000	
Selling price of land	30,000	
Selling price of machinery	25,000	
Working capital generated from other sources		65,000
Total working capital generated		$ 95,000
Financial resources generated not affecting working capital:		
Bonds issued for land acquired	$25,000	
Preferred stock issued to retire bonds payable	20,000	
Common stock issued to acquire patent	10,000	
Total		55,000
Total financial resources generated		$150,000
Financial resources applied:		
Working capital applied:		
Cash dividend paid	$20,000	
Land purchased	9,000	
Working capital applied		$ 29,000
Financial resources applied not affecting working capital:		
Land acquired by issuing bonds payable	$25,000	
Bonds payable retired by issuing preferred stock	20,000	
Patent acquired by issuing common stock	10,000	
Financial resources applied not affecting working capital		55,000
Increase (decrease) in net working capital during the period		66,000
Total financial resources applied		$150,000

XYZ Company
Income Statement
for the period January 1, 1974 to December 31, 1974

Gross sales		$202,600
Less: Sales return		8,000
Net sales		$194,600
Cost of goods sold (including depreciation expense of $5,000)		75,000
Gross profit on sales		$119,600
Operating expenses:		
Selling expenses (including depreciation expense of $4,500)	$36,500	
Administrative expenses	32,000	
General expenses (including amortization of patent $400)	14,500	$ 83,000
Operating income		$ 36,600
Other income:		
Gain on sale of land	$ 1,000	
Gain on sale of machinery	5,000	
	$ 6,000	
Other expenses:		
Interest expenses (including amortization of bond discount $100)	1,600	
Net other income and expense		4,400
Income before income taxes		$ 41,000
Federal and state income taxes		15,000
Net income		$ 26,000

XYZ Company
Balance Sheet
at December 31, 1973 and 1974

ASSETS	1974		1973	
Current Assets				
Cash		$ 86,000		$ 50,000
Marketable securities		5,000		10,000
Accounts receivable	$ 85,000		$ 68,000	
Less: Allowance for uncollectible accounts	10,000	75,000	8,000	60,000

		1974		1973
Inventory		30,000		20,000
Prepaid expenses		5,000		2,000
Total Current Assets		$201,000		$142,000

Property, Plant, and Equipment

Land		$ 40,000		$ 35,000
Building	$200,000		$200,000	
Less: Accumulated depreciation	65,000	135,000	60,000	140,000
Machinery and equipment	$ 80,000		$110,000	
Less: Accumulated depreciation	54,500	25,500	60,000	50,000
Total Net Property, Plant, and Equipment		$200,500		$225,000

Intangible Assets

Patent		$ 25,600		$ 16,000
TOTAL ASSETS		$427,100		$383,000

EQUITIES

Current Liabilities

Accounts payable		$ 70,000		$ 90,000
Notes payable		5,000		7,000
Income taxes payable		15,000		
Total Current Liabilities		$ 90,000		$ 97,000

Long-Term Liabilities

Bond payable	$ 35,000		$ 25,000	
Less: Bond discount	400	34,600	500	24,500
Total Liabilities		$124,600		$121,500

Owners' Equity

Preferred stock		$ 70,000		$ 50,000
Common stock		115,000		100,000
Capital in excess of par value — common stock		14,500		14,500
Retained earnings		103,000		97,000
		$302,500		$261,500
TOTAL LIABILITIES AND OWNERS' EQUITY		$427,100		$383,000

The objective of a statement of changes in financial position prepared on a working capital basis, utilizing the all financial resources concept, is to explain the change in working capital by explaining *all* changes in the noncurrent accounts that have occurred during the year. To explain these changes, the person preparing the statement must be aware of all entries that have affected the accounts. When the statement is completed, the change in each noncurrent account will have been explained. Therefore, the preparation of this statement requires an understanding of the accounting procedures described in the preceding chapters. A review of the earlier material relative to the entries for stock dividends, sale of fixed assets, and other unusual transactions may be helpful in understanding this statement.

Even though a purpose of the statement is to explain the changes in working capital, the statement includes financing activities which do not affect working capital in order to present a more complete explanation of the changes in financial position. Thus, in a broad context, the term *flow of funds* refers both to the flow of working capital and the flow of noncurrent resources which do not affect working capital. In the following discussion, the term *flow of funds* will refer both to the working capital and financial resources.

Determination of Working Capital

The initial calculation involves the determination of the change in working capital from the beginning of the period to the end of the period. Working capital has been defined as *the excess of current assets over current liabilities*.

In the example, the working capital was calculated as:

	12/31/74	12/31/73
Current assets	$201,000	$142,000
Current liabilities	90,000	97,000
Working capital	$111,000	$ 45,000

Thus, the change in the working capital is an increase of $66,000 ($111,000 − $45,000). There has been an excess inflow of funds over the outflow of funds in the amount of $66,000. In the final preparation of the statement, there will be an excess of resources generated over resources applied in the amount of the change in the working capital. Therefore, by determining the change in the working capital as the first step, the amount to be explained by the statement (difference between resources generated and applied) is known in advance. This knowledge may be helpful in the preparation of the statement.

The change in working capital, in the previous example, was calculated by adding the current assets and subtracting the current liabilities. An alternative calculation would be to determine the change in the individual components of current assets and current liabilities.

XYZ Company
Schedule of Changes in Working Capital

	Account balance		Working capital increase (decrease)
Changes in working capital accounts:	12/31/74	12/31/73	
Current Assets:			
Cash	$ 86,000	$ 50,000	$36,000
Marketable securities	5,000	10,000	(5,000)
Accounts receivable (net)	75,000	60,000	15,000
Inventory	30,000	20,000	10,000
Prepaid expenses	5,000	2,000	3,000
Total current assets	$201,000	$142,000	
Current Liabilities:			
Accounts payable	$ 70,000	$ 90,000	$20,000
Notes payable	5,000	7,000	2,000
Income taxes payable	15,000		(15,000)
Total current liabilities	$ 90,000	$ 97,000	
Working Capital	$111,000	$ 45,000	$66,000

The schedule of changes in working capital may be summarized as follows: *increases in working capital result from either increases in current assets or decreases in current liabilities. Decreases in working capital are attributable to decreases in current assets or increases in current liabilities.* These changes can be seen from the initial calculation of working capital (current assets minus current liabilities).

The following example summarizes the effect on working capital of the changes in current assets and current liabilities:

	Case 1	Case 2	Case 3	Case 4
Current assets	Increase	Decrease	Same	Same
− Current liabilities	Same	Same	Increase	Decrease
Working capital	Increase	Decrease	Decrease	Increase

In each case, only the assets or liabilities were changed. In the event that both change or two or more accounts change (as in the example), the effect on working capital can be determined only by noting the sum of the individual changes. In the example of XYZ Company, the increase of $66,000 resulted primarily from the $36,000 increase in cash and the $20,000 decrease in accounts payable. This type of information could not be obtained by subtracting total current liabilities from total current assets.

Now that the change in working capital has been determined to be an increase of $66,000, the next question is: What caused the increase? The only items that could produce the change in the working capital are those items that *are not* part of working capital. This means that investments, fixed assets, intangibles, long-term liabilities, and stockholders' equity are the only items that could produce the change. The reason for the noncurrent accounts producing the change in the current accounts is simple if viewed in terms of the balance sheet equation. The balance sheet must always have an equality of assets and equities. Therefore, the removal of the current asset and current liability amounts, which differ by the amount of the working capital, necessarily means that the non-current accounts differ by the same amount.

ADJUSTMENT OF INCOME STATEMENT TO A FUNDS STATEMENT

Once the change in working capital has been determined, the next step in preparing the funds statement is the adjustment of the income statement. The income statement must be adjusted because some revenues and expenses do not involve flows of funds. For example, depreciation does not involve an outflow of working capital, but it is an expense. Depreciation is the allocation of the cost of the asset, which was an outflow at the time of purchase, to the periods benefitted by that asset. The entry for depreciation clearly demonstrates that the amount of depreciation recorded in any period has no effect on working capital or cash.

| Depreciation Expense | 1,000 | |
| Accumulated Depreciation | | 1,000 |

Other items that are additions or deductions from revenue but do not involve flows of funds are:

1. Amortization of intangibles. The amortization of an intangible has the same effect as the depreciation expense. The amount of the write-off reduces the intangible asset but has no effect on the flow of funds.

2. Amortization of bond discount or premium. This entry has no effect on the amount of the interest that is paid or received. The amortization does not involve a flow of funds. The flow of funds is the amount of interest received or paid.

3. Gains or losses on the sale of fixed assets. These represent the difference between the selling price and book value. The flow of funds is the selling price and not the gain or loss.

The adjustment of the income statement to a funds flow statement is accomplished by adding back those expenses that did not require outflows of funds and by deducting those revenues that were not inflows of funds. As shown in the example:

Net income		$26,000
Add (deduct) items not requiring or generating working capital in the current period:		
Depreciation expense	$9,500	
Amortization of bond discount	100	
Amortization of patent	400	
Gain on sale of land	(1,000)	
Gain on sale of machinery	(5,000)	4,000
Working capital generated by operations		$30,000

Evaluation of Noncurrent Balance Sheet Accounts

The other changes in funds can be determined through the use of the T account method. Using this approach, the beginning and ending balances of each account are placed in a T account. The transactions which produced the change are then placed in the account. As a general rule, the first account to be analyzed is the first noncurrent asset. In the example, the first noncurrent account is Land. The remainder of the chapter will be based upon the financial statements of XYZ Company.

Land

The T account for Land has an increase of $5,000, which resulted from three separate transactions. These transactions are:

LAND

Beginning balance	35,000		
Land purchased for cash	9,000	29,000	Cost of land sold
Land received in exchange for bonds	25,000		
Ending balance	40,000		

The purchase of land is an outflow of funds while the sale of land represents a source of funds. Thus, in the illustration, the selling price of the land was a source of $30,000 and appears under the caption, Other Sources of Working Capital. The difference between the selling price and the cost ($30,000 − $29,000) was a gain of $1,000 and appears on the income statement as Other Income.

The land purchased for $9,000 is a use of funds and appears under the caption, Working Capital Applied. Cash was expended to buy the land; thus this was an outflow of working capital. In addition, land was acquired through the issuance of bonds. This transaction did not affect working capital but is a part of the over-all financing activities of the firm. Consequently, through the use of the all financial resources concept, this transaction should appear on the funds statement under the caption, Financial Resources Applied Not Affecting Working Capital.

Buildings, Machinery, and Equipment

These accounts may produce problems because, in addition to the cost and selling price of these items, the accumulated depreciation and depreciation expense are also involved. The purchase of fixed assets has the same effect as the purchase of land.

The purchase of fixed assets for cash or other working capital items is a use of funds and would appear under the caption, Working Capital Applied. The purchase through the issuance of long-term bonds or stock would be placed under the caption, Financial Resources Applied Not Affecting Working Capital.

The sale of these assets requires a thorough investigation, not only of the asset account but also of the accumulated depreciation account. This is necessary because the entry to sell the asset involves the following accounts:

Cash (amount received)
Accumulated depreciation (to date)

Asset (original cost)

Loss or gain on sale (difference between book value and selling price)

The cash received is the source of funds irrespective of the book value or gain or loss on the sale. The book value represents the undepreciated amount on the books and the gain or loss represents the difference between the book value and the economic value at the time of sale. The debit to accumulated depreciation must be known so that the remaining change in the account can be explained by the depreciation expense for the year. In the example, the Accumulated Depreciation—Machinery account was:

ACCUMULATED DEPRECIATION—MACHINERY

	60,000	Beginning balance
Sale of machinery 10,000 ($30,000 − $10,000 = book value $20,000)	4,500	Depreciation expense for the year
	54,500	Ending balance

The $10,000 debit is the accumulated depreciation on the machinery which was sold for $25,000. The $4,500 credit is the result of the depreciation expense entry for the year. The sale and depreciation expense for the year thus explain the decrease of $5,500 in the account.

The Accumulated Depreciation—Buildings account is somewhat easier to explain. There were no sales or purchases of buildings during the year, and the change in the account is the result of the depreciation expense for the year. This is shown as follows:

ACCUMULATED DEPRECIATION—BUILDINGS

	60,000	Beginning balance
	5,000	Depreciation expense for the year
	65,000	Ending balance

Therefore, to explain the changes in the accumulated depreciation accounts, the entries made for any sales of the related assets must be known and the depreciation expense for the year must also be determined.

The fixed asset accounts also must be reviewed in order to determine if any assets were purchased during the year. In the example, the Buildings account did not change during the year. The Machinery and Equipment account decreased by $30,000 and must be explained. The reason for the decrease is the sale of machinery (original cost $30,000) for $25,000. The effect on the T account is shown as follows:

MACHINERY

Beginning balance	110,000		
		30,000	Cost of machinery sold
Ending balance	80,000		

The source of funds is the selling price of the machinery, $25,000. Since the book value was $20,000 (Original Cost $30,000 — Accumulated Depreciation $10,000), the difference of $5,000 was reported on the income statement as a gain.

Intangibles

Purchases of intangible assets represent a use of funds and would be shown under the caption, Working Capital Applied. The purchase of a patent through the issuance of long-term debt or stock does not require the use of working capital. Therefore, it would appear under the caption, Financial Resources Applied Not Affecting Working Capital.

The selling price of the intangible is a source of funds. This discussion is not intended to be applicable only to patents since all intangibles are to be analyzed in the same manner. The change in the Patent account is explained as follows:

PATENTS

Beginning balance	16,000		
Acquired through issuance of common stock	10,000	400	Amortization for the year
Ending balance	25,600		

Note that the amortization for the year is a direct reduction of the asset account. Normally an allowance account is not created for an intangible asset. Consequently, the amount of amortization must be determined from the income statement in order to completely explain the change in the account.

In summary, the following points should be noted about all non-current assets:

1. *The selling price of the asset is the source of funds.*
2. *The amount purchased through the use of working capital is shown under the caption, Working Capital Applied.*
3. *Purchases made through the issuance of long-term debt or stock do not require the use of working capital but must be shown on the funds statement. These purchases are shown under the caption, Financial Resources Applied Not Affecting Working Capital.*

Long-Term Liabilities

The liabilities have the opposite effect on working capital from the assets. The creation of a liability is a source of working capital, while a reduction is a use of funds. The reason for this is that the creation of a liability means that cash, or other assets, was not needed to pay for the item purchased. Also, the creation of a liability may involve an inflow of cash. For example, the sale of a bond issue results in a cash inflow into the company and the creation of a liability for future repayment. The inflow of cash is a source of funds in the present period, while the repayment of the debt will be an outflow of funds in some future period or periods.

Long-term debt created for the acquisition of investments, fixed assets, or intangible assets is also a source of financial resources. Since the noncurrent assets do not affect working capital, the issuance of long-term debt is placed under the caption, Financial Resources Generated Not Affecting Working Capital.

As discussed, the amortization of bond discount or premium is an adjustment of net income. This information must also be used in determining the changes in the net amount of long-term debt. In the example, the change from $24,500 to $34,600 can be explained as follows:

BONDS PAYABLE

		25,000	Beginning balance
		5,000	Sale of bond
Retired through the issuance of preferred stock	20,000	25,000	Issued for land acquired
		35,000	Ending balance

DISCOUNT ON BONDS PAYABLE

Beginning balance	500		
		100	Amortization for the year
Ending balance	400		

Note that the sources of funds were the selling price of bonds (at par), $5,000, and the issuance of bonds for land, $25,000. The retirement of the bonds through the issuance of the preferred stock is a use of funds. Since the retirement did not affect working capital, the use is recorded under the caption, Financial Resources Applied Not Affecting Working Capital. The amortization of the discount is added to net income since no funds were required.

Owners' Equity

Owners' Equity accounts are similar to liabilities in that increases are, generally, sources of funds and decreases are uses of funds. However, there are situations, e.g., stock dividends, wherein there is merely a rearrangement of the owners' equity accounts, in which case there is neither a source nor use of funds.

Preferred Stock. The issuance of preferred stock either for cash or other assets is a source of funds. If the stock is sold for cash or working capital, the amount appears under the caption, Other Sources of Working Capital. If the stock is issued for assets other than working capital, the source appears under the caption, Financial Resources Generated Not Affecting Working Capital. The amount of cash received, or the value of the assets, is the source of funds. The par value of the stock is not relevant in the determination of the funds provided. If a preferred stock, par $100, was sold for $110, the source of funds is the sale price, not the par value. The excess over the par value would be verified by the change in the Capital in Excess of Par Value—Preferred Stock account.

The change in the Preferred Stock account is illustrated as follows:

PREFERRED STOCK

	50,000	Beginning balance
	20,000	Issued to retire bonds payable
	70,000	Ending balance

Common Stock. The guidelines for preferred stock are equally applicable to common stock. The increase in common stock is normally a source of funds while the decrease is a use of funds.

Stock dividends are an exception to the general rule that increases in owners' equity accounts are sources of funds. As stated in Chapter 11, a stock dividend requires the capitalization of retained earnings. As such, the *only* result of the transaction is the transferral from Retained Earnings to the Common Stock and Paid-In Capital accounts of an amount equal to the market value of the stock times the number of shares. Any changes in the accounts must be analyzed, but any change resulting from a stock dividend should be ignored for fund statement purposes. This is also true for a stock split although most splits do not require journal entries; therefore, the accounts would not be affected.

In the example, the change of $15,000 in the Common Stock account can be analyzed as follows:

COMMON STOCK

	100,000	Beginning balance
	5,000	Sale at par
	10,000	Issued to acquire patent
	115,000	Ending balance

Note that the Capital in Excess of Par Value—Common Stock account did not change. This is an indication that any transactions affecting common stock took place at par value.

Retained Earnings. This account is affected by many of the items that have been previously mentioned, i.e., stock dividends and net income. In addition, cash dividends will affect this account. Consequently, the various changes must be differentiated so that those affecting working capital can be determined.

The increase of $6,000 for the year in retained earnings is accounted for as follows:

RETAINED EARNINGS

		97,000	Beginning balance
		26,000	Net income
Cash dividend declared	20,000		
		103,000	Ending balance

Note that the declaration of cash dividends is a use of funds even though cash was not distributed. The fact that a current liability was created results in a reduction of working capital. The actual payment of the dividend does not affect working capital since both current assets and liabilities are reduced. Thus the declaration, not the payment of a cash dividend, is the amount shown under the caption, Working Capital Applied.

As a final comment on the Owners' Equity section, treasury stock has the same effect on working capital as does the purchase of an asset. That is, the purchase of treasury stock is a use of funds and the selling price of the treasury stock is a source of funds. Since treasury stock is the last item to be explained, every change on the balance sheet has now been analyzed.

In summary, the following points should be noted about the long-term liabilities and owners' equity:

1. *Increases, normally, mean a source of funds. The cash received or the value of the asset is the amount of the source.*

2. *Decreases, normally, mean a use of funds. The cash paid or value given up is the amount of the use of funds.*

3. *A stock dividend is a rearrangement of the Stockholders' Equity section and is ignored for fund purposes.*

STATEMENT OF CHANGES IN FINANCIAL POSITION — CASH BASIS

A statement of changes in financial position prepared on a cash basis is the historical counterpart of the cash budget which was discussed in Chapter 5. The purpose of the statement, prepared on a cash basis, is to provide the details relative to cash generated and applied during the period. The difference between the amount generated and the amount applied is, obviously, the change in cash from the beginning of the period to the end of the period. When the historical statement is related to the forecasted statement, management and investors are provided information for the evaluation of past performance and for planning the future activities of the firm.

Format

The procedures for the preparation of the statement of changes on a cash basis are basically the same as for the working capital basis. There are certain differences, however, because the change in cash is to be explained in contrast to the change in working capital.

The format of the statement is the same except for the addition of the new caption, Changes in Current Assets and Liabilities Affecting Cash. This caption is necessary in order to include the effect of the other current accounts on the cash account. If the statement for XYZ Company had been prepared on a cash basis, this new caption would appear immediately below Working Capital Generated by Operations. The information would be presented as follows:

Changes in current assets and
liabilities affecting cash:

Accounts receivable	Increase	$(15,000)
Marketable securities	Decrease	5,000
Inventory	Increase	(10,000)
Prepaid expenses	Increase	(3,000)
Accounts payable	Decrease	(20,000)
Notes payable	Decrease	(2,000)
Income taxes payable	Increase	15,000
Changes in current accounts		$(30,000)

The changes in current accounts, together with the adjustment of net income, have the effect of reporting operations on the basis of cash flow. For example, sales on account, which increase net income, do not provide any cash flow until the receivable is collected. Thus the increase in accounts receivable is considered to be a use of cash.

Cash is affected by changes in the current assets and liabilities in the same manner as working capital is affected by noncurrent items. A decrease in a current asset is a source of funds because cash is provided, e.g., collection of a receivable results in an inflow of cash, and a decrease in prepaid expenses means that cash was not used to pay the expense in the current period. An increase of a current asset is a use of cash, e.g., purchase of inventory reduces cash.

An increase in current liabilities is considered to be a source of funds because cash is being provided either by a cash borrowing or the elimination of the need for cash to be paid. In the example, note that the increase of the accounts receivables, inventories, and prepaid expenses were considered uses of cash under the caption, Changes in Current Assets and Liabilities Affecting Cash. In addition, the decrease of the trade payable was also a use of funds. On the other hand, the decrease in the marketable securities was a source of funds because cash was made available through their sale.

In summary, the points to be noted concerning current assets and liabilities are:

1. *Decreases in current assets and increases in current liabilities are considered sources of cash.*
2. *Increases in current assets and decreases in current liabilities are considered uses of cash.*

SUMMARY

The statement of changes in financial position is considered essential for statement users, especially owners and creditors, in making economic decisions. The purpose of the statement is to supplement, not replace, the income statement and balance sheet by providing details of the changes in financial position during the period.

There are several definitions of the term funds, with the more common ones being (a) cash, (b) working capital, and (c) all financial resources. The recommended definition is that of all financial resources. Accordingly, the statement of changes in financial position should include the details of all financial resources even though the statement may be prepared on either a cash or working capital basis.

The difference between the cash and working capital basis is in the treatment of current assets and liabilities. In a statement prepared on the working capital basis, only the net change in working capital is explained. Under the cash basis each of the current accounts, other than cash, as well as the noncurrent accounts, must be explained in order to develop the change in cash.

The statement of changes in financial position reflects the financial resources generated and applied during the period.

Financial resources are generated by:

1. Operations.
2. Sale of noncurrent assets.
3. Long-term borrowing.
4. Owners' investment.
5. Inflows which do not affect cash or working capital, e.g., issuance of bonds for fixed assets.

Financial resources are applied to:

1. Purchase of noncurrent assets.
2. Payment of long-term borrowing.
3. Payment of dividends.
4. Outflows which do not affect cash or working capital, e.g., fixed assets acquired through issuance of common stock.

QUESTIONS

12-1. What, in general, is the purpose of a statement of changes in financial position?

12-2. What is working capital?

12-3. What are financial resources?

12-4. What is the purpose of a statement of changes in financial position when prepared on a working capital basis?

12-5. Explain the three basic concepts of funds.

12-6. What is the purpose of a statement of changes in financial position when prepared on a cash basis?

12-7. How does "working capital provided by operations" differ from net income?

12-8. In a statement prepared on a working capital basis, why is it important to begin with the change in working capital?

12-9. Why do the noncurrent accounts explain the change in the current accounts?

12-10. Will cash generated by operations be the same as working capital generated by operations? Explain.

12-11. Answer the following questions assuming that land costing $15,000 is sold for $20,000:
(a) How much working capital is provided?
(b) Under what caption on the funds statement would this item appear?

PROBLEMS

Procedural Problems

12-1. For each of the following transactions, prepare the appropriate journal entry and indicate whether cash and working capital are increased, decreased, or not affected.
(a) Paid employees accrued salaries, $2,000.
(b) Purchased merchandise on credit, $5,000.
(c) Depreciation adjusting entry, $1,200.
(d) Declared dividends, $1,000.
(e) Paid for merchandise purchased on credit, $3,000.
(f) Amortization of premium on bonds payable, $200.
(g) Sold marketable securities costing $1,200 for $1,500.
(h) Sold common stock, $2,000.

12-2. The income statement of Howard Supply Company showed a net loss of $10,000. Included on the statement was a deduction for depreciation expense of $12,000, amortization of intangibles of $4,000, and amortization of bond premium of $2,000.

REQUIRED
(a) Compute the working capital provided by operations.
(b) If the answer in (a) is different from the net loss, explain the reason for the difference.

12-3. The balance sheets for Holiday Corporation contained the following information:

	December 31	
	1975	1974
Cash	5,000	6,000
Accounts receivable	8,000	4,000
Inventory	15,000	10,000
Investments		2,000
Fixed assets	47,000	35,000
	$75,000	$57,000
Allowance for uncollectible accounts	$ 1,000	$ 500
Accumulated depreciation on fixed assets	7,000	5,000
Accounts payable	4,000	3,000
Notes payable — current	7,000	4,000
Notes payable — long-term	20,000	15,000
Common stock	25,000	25,000
Retained earnings	11,000	4,500
	$75,000	$57,000

Additional information concerning changes in the noncurrent accounts:
• Net income for 1975 was $6,500.
• Depreciation on the fixed assets was $2,000.
• Investments were sold at cost.
• Purchased fixed assets costing $12,000; paid cash of $7,000 and a long-term note of $5,000.

REQUIRED
(a) Determine the change in working capital.
(b) Prepare a statement of changes in financial position (working capital basis).

12-4. Using the data in Problem 12-3:
(a) Prepare a statement of changes in financial position (cash basis).
(b) Reconcile the results on the cash basis with the statement prepared in Problem 12-3 on the working capital basis.

12-5. The following balance sheet, income statement, and supplemental notes are provided for ABC Company.
REQUIRED. Prepare a statement of changes in financial position on the working capital basis using all financial resources for 1974.

ABC Company
Balance Sheet
at December 31, 1973 and 1974

ASSETS		1974		1973
Current Assets				
Cash		$ 145,000		$ 100,000
Marketable securities		55,000		150,000
Accounts receivable	$400,000		$425,000	
Less: Allowance for uncollectible accounts	20,000	380,000	25,000	400,000
Notes receivable		10,000		8,000
Inventories		462,000		454,000
Total Current Assets		$1,052,000		$1,112,000
Investments				
Investment in long-term securities		$ 90,000		$ 100,000
Land held for future plant site		–		50,000
Total investments		$ 90,000		$ 150,000
Property, Plant, and Equipment				
Land		$ 150,000		$ 100,000
Buildings	$825,000		$715,000	
Less: Accumulated depreciation	250,000	575,000	230,000	485,000
Machinery and equipment	$550,000		$450,000	
Less: Accumulated depreciation	90,000	460,000	75,000	375,000
Total Property, Plant, and Equipment		$1,185,000		$ 960,000
Intangibles and Other Assets				
Patents		$ 14,000		$ 10,000
Copyright		5,000		6,000
Total Intangible and Other Assets		$ 19,000		$ 16,000
TOTAL ASSETS		$2,346,000		$2,238,000

PROBLEMS

EQUITIES	1974	1973
Current Liabilities		
Accounts payable	$ 260,000	$ 210,000
Notes payable (due in 1 year)	10,000	40,000
Income tax payable	143,000	103,000
Sales commission payable	28,000	26,000
Total current liabilities	$ 441,000	$ 379,000
Long-Term Liabilities		
Bonds payable	$ 100,000	$ 200,000
Total Liabilities	$ 541,000	$ 579,000
Owners' Equity		
Common stock, $10 par value authorized shares 60,000, issued shares 54,000	$ 540,000	$ 500,000
Premium on common stock	154,000	150,000
Retained earnings	1,111,000	1,009,000
Total Owners' Equity	$1,805,000	$1,659,000
TOTAL LIABILITIES AND OWNERS' EQUITY	$2,346,000	$2,238,000

ABC Company
Income Statement
for the Year 1974

Net sales		$4,782,000
Less: Cost of goods sold		3,464,000
Gross profit on sales		$1,318,000
Operating expenses:		
Selling expenses	$288,000	
Administrative expenses	330,000	
General expenses	429,000	1,047,000
Operating income		$ 271,000
Other income		
Gain on sale of building	$ 14,000	
Gain on sale of investment	2,000	16,000
Net income before taxes		$ 287,000
Federal and state income taxes		135,000
Net income		$ 152,000

Supplemental Notes:

- A fully depreciated building was sold (net of tax) for $14,000, original cost $100,000.
- A new building costing $210,000 was acquired through the issuance of 4,000 shares of stock and the payment of $166,000 cash.
- A long-term investment which cost $10,000 was sold (net of tax) for $12,000.
- Land which cost $50,000 was converted from investments into a building site for the new building.
- Machinery costing $100,000 was purchased for cash.
- A new patent was purchased for $4,000.
- Bonds payable in the amount of $100,000 were redeemed.
- Dividends in the amount of $50,000 were paid to the common stockholders.

12-6. Using the information provided in Problem 12-5, prepare a statement of changes in financial position on the cash basis using all financial resources for 1974.

12-7. The following balance sheet, income statement, and supplemental notes are provided for Williams Corporation.

REQUIRED. Prepare a statement of changes in financial position on the working capital basis using all financial resources for 1975.

Williams Corporation
Balance Sheet
at December 31, 1974 and 1975

ASSETS	1975		1974	
Current Assets				
Cash		$ 33,500		$ 27,000
Accounts Receivable	$ 91,800		$ 81,700	
Less: Allowance for uncollectible accounts	1,900	89,900	2,000	79,700
Inventories		136,300		133,200
Prepaid expenses		4,600		12,900
Total Current Assets		$264,300		$252,800
Investments				
Land held for future plant site		$ 35,000		

	1975		1974	
Property, Plant, and Equipment				
Land		$ 47,000		$ 47,000
Buildings	$463,000		$280,000	
Less: Accumulated depreciation	93,100	369,900	65,000	215,000
Machinery and equipment	$244,500		$262,000	
Less: Accumulated depreciation	62,500	182,000	52,000	210,000
Total Property, Plant, and Equipment		$598,900		$472,000
Intangibles and Other Assets				
Patents		$ 1,000		$ 2,000
Organization cost		500		1,000
Total Intangible and Other Assets		$ 1,500		$ 3,000
TOTAL ASSETS		$899,700		$727,800

EQUITIES

	1975	1974
Current liabilities		
Accounts payable	$ 3,000	$ 7,800
Notes payable	8,000	5,000
Mortgage payable	3,600	3,600
Accrued liabilities	6,200	4,800
Income tax payable	87,500	77,900
Total Current Liabilities	$108,300	$ 99,100
Long-Term Liabilities		
Notes payable		$ 18,000
Mortgage payable	$ 70,200	73,800
Total Long-Term Liabilities	$ 70,200	$ 91,800
Deferred Income Tax	$ 16,800	$ 18,900
Total Liabilities	$195,300	$209,800
Owners' Equity		
Common stock—$1 par value, authorized shares, 300,000; issued shares, 162,000 in 1975 and 120,000 in 1974	$162,000	$120,000

	1975	1974
Capital in excess of par value — common	306,900	197,900
Retained earnings	235,500	200,100
Total Owners' Equity	$704,400	$518,000
TOTAL LIABILITIES AND OWNERS' EQUITY	$899,700	$727,800

Williams Corporation
Income Statement
for the Year 1975

Net sales		$974,000
Less: Cost of goods sold		540,000
Gross profit on sales		$434,000
Operating expenses:		
Selling expenses	$ 52,000	
Administrative expenses	87,000	
General expenses	123,000	262,000
Operating income		$172,000
Other income and expense		
Gain on sale of machinery		3,000
Net income before taxes		$175,000
Federal and state income taxes		85,400
Net income		$ 89,600

Supplemental Notes:
- A 10% stock dividend was distributed on August 1 when the fair market value of the stock was $3 per share.
- The investment in land for a future plant site was obtained by the issuance of 10,000 shares of the Corporation's common stock on October 1.
- On December 1, the firm sold 20,000 shares of common stock at $4 per share to obtain additional working capital.
- During 1975, fully depreciated machinery which cost $17,500 was sold for $3,000.
- The only depreciable asset acquired in 1975 was a new building which was completed in December; no depreciation was taken on its cost.

- In 1975, $10,000 was paid in advance on long-term notes payable. The balance of the long-term notes is due in 1976.
- Cash dividends in the amount of $18,200 were paid to the common stockholders.

12-8. Using the information provided in Problem 12-7, prepare a statement of changes in financial position on the cash basis using all financial resources for 1975.

12-9. The balance sheets of the Ross Corporation for the years ended December 31, 1974 and 1975 are presented as follows:

	December 31	
	1975	1974
Cash	$ 8,000	$ 6,000
Accounts receivable	12,000	7,000
Prepaid rent	2,400	3,600
Inventory	28,000	32,000
Long-term investments	40,000	40,000
Fixed assets	88,000	70,000
	$178,400	$158,600
Accumulated depreciation on fixed assets	$ 21,500	$ 20,000
Accounts payable	9,000	7,200
Taxes payable	6,000	6,000
Notes payable — short-term	26,000	14,000
Mortgage payable	35,000	35,000
Common stock	60,000	50,000
Retained earnings	20,900	26,400
	$178,400	$158,600

Supplemental Notes:
- Purchased a new fixed asset costing $25,000. Paid $13,000 cash and gave a short-term note for the remainder.
- Sold $10,000 of common stock at par value.
- Disposed of a fully depreciated asset having an original cost of $7,000 and no salvage value.
- Net loss for the year amounted to $1,500.
- Dividends paid during 1975 amounted to $4,000.

REQUIRED
(a) Compute the change in working capital for the year.
(b) Prepare a statement of changes in financial position (working capital basis).

12-10. The following account balances represent the year-end trial balance amounts of Lovell Corporation:

	December 31 1975	December 31 1974	Increase or (decrease)
Cash	$ 21,500	$ 13,000	$ 8,500
Receivables, net	57,600	36,100	21,500
Inventories	84,300	81,500	2,800
Investments (long-term)	60,000	70,000	(10,000)
Fixed assets	215,000	165,000	50,000
Land	100,000	100,000	–0–
Accumulated depreciation	56,000	38,000	18,000
Accounts payable	36,400	30,500	5,900
Salaries payable	5,000	7,000	(2,000)
Mortgage payable	100,000	75,000	25,000
Common stock	200,000	200,000	–0–
Retained earnings (January 1)	115,100	81,300	33,800
Sales	168,500	142,700	25,800
Other income	12,600	15,000	(2,400)
Cost of goods sold	76,200	68,900	7,300
Selling and administrative expense	24,000	26,000	(2,000)
Depreciation expense	18,000	14,000	4,000
Interest expense	5,000	5,000	–0–
Loss on sale of investments	2,000	–0–	2,000
Dividends	30,000	10,000	20,000

REQUIRED
(a) Prepare a statement of retained earnings beginning with the January 1, 1974 balance and show the total retained earnings as of the end of 1974 and the end of 1975.
(b) Compute the change in working capital.
(c) Prepare a statement of change in financial position (working capital basis). The following additional information concerns changes in the noncurrent accounts:
   (1) Investments were sold early in 1975.
   (2) Purchased fixed assets paying half in cash and assuming a mortgage for the other half.

12-11. The records of the Yalc Company included the following information relative to the balance sheet accounts:

| | December 31 | |
Debits	1974	1973
Cash	$ 27,800	$ 19,600
Accounts receivable, net	41,500	36,900
Marketable securities	21,000	17,000
Inventory	78,300	69,800
Prepaid insurance	6,000	–0–
Long-term investments	54,800	62,500
Building	120,000	150,000
Machinery	38,000	38,000
Goodwill	12,000	15,000
	$399,400	$408,800

| | | |
Credits		
Accumulated depreciation — building	$ 20,000	$ 48,000
Accumulated depreciation — machinery	15,000	12,000
Accounts payable	26,500	25,400
Notes payable — short-term	18,600	15,000
Accrued salaries	7,000	9,000
Taxes payable	12,000	12,000
Notes payable — long-term	55,000	60,000
Mortgage payable	75,000	75,000
Common stock	100,000	85,000
Retained earnings	70,300	67,400
	$399,400	$408,800

Supplemental Notes:
- The long-term investments were sold for $10,000.
- Sold a fully depreciated building which originally cost $50,000 for $12,000. The sale price was $2,000 greater than the building's salvage value.
- Goodwill is to be amortized over a five-year period beginning in 1973.
- A new building was purchased with 75% of the purchase price being paid with common stock issued at par value. The remainder was paid in cash.
- Net income for the year amounted to $30,900.
- A cash dividend was declared and paid in December.

REQUIRED
(a) Compute the change in working capital for the year.
(b) Prepare a statement of changes in financial position on the working capital basis using all financial resources for 1974.

12-12. Using the information provided in Problem 12-11, prepare a statement of changes in financial position on the cash basis using all financial resources for 1974.

13

FINANCIAL STATEMENT ANALYSIS

EDUCATIONAL OBJECTIVES

The material in this chapter is designed to achieve several educational objectives. These include:

1. An understanding of the importance of comparative financial statements in the evaluation of the financial stability and prospects of a business.

2. A knowledge of the technical procedures required for the calculation of various ratios and how those ratios may be used in evaluating the financial statements of a business.

3. An appreciation of the fact that financial ratios are the result of using the dollar amounts in the financial statements; therefore, the usefulness of the ratios depends upon the validity of the numbers in the financial statements.

4. A knowledge that footnotes are an integral part of the financial statements and should be included in any financial analysis of a business.

5. An awareness that dollar amounts on the financial statements may be converted into percentages, thus enhancing the comparability of different size firms.

The purpose of accounting is to provide quantitative financial information as an input to the decision-making process. In achieving this purpose, accounting involves the processes of identifying, measuring, and communicating useful economic information to those who must make economic decisions. The preceding chapters of this book contained both discussions and illustrations of the rationale and procedures used by accountants in the identification, measurement, and communication processes. There is at least one other process necessary before the user of accounting information can make a decision, and that is an interpretation of the information.

The analysis, or interpretation, of financial statements is sometimes viewed as the final stage in the accounting process. Others view the analysis as the responsibility of the user and not the accountant. The latter view is perhaps more appropriate inasmuch as there are numerous users, each of whom may have a different opinion as to what is the most useful analysis. In any event, the analysis generally takes the form of calculations based upon the quantitative information provided in the financial statements.

The purpose of this chapter is to present several of the more common calculations along with an explanation of their significance to the reader of the statements. These calculations should not be presumed to be the only analyses which may be used. The variety of analyses is limited only by the desire of the individual using the statements; however, the ones presented are fairly common and serve to illustrate the possibility for financial statement analysis.

COMPARATIVE STATEMENTS

Many companies present comparative financial statements for at least two years, and statements for five to ten years are not uncommon. The value of such comparative statements has long been realized by accountants, as well as others. Accordingly, the Committee on Accounting Procedure of the AICPA expressed a strong case for comparative statements. In *Accounting Research Bulletin 43*, the Committee stated:

> The presentation of comparative financial statements in annual and other reports enhances the usefulness of such reports and brings out more clearly the nature and trends of current changes affecting the enterprise. Such presentation emphasizes the fact that statements for a series of periods are far more significant than those for a single period and that the accounts for one period are but an installment of what is essentially a continuous history.

Format of Comparative Statements

In this chapter, only two years of comparative statements will be illustrated in order to conserve space and because many companies only present the current and preceding years. The statement example used in this chapter will be the Widget Company illustration which was presented in Chapter 2. In addition to the basic information for each of the two years, some companies have a column for increases and decreases of the current year over the previous year.

The Widget Manufacturing Company, Inc.
Balance Sheet
at December 31, 1974
with Comparative Figures for 1973

		1974		1973	Increase (decrease) 1974 over 1973
Current Assets					
Cash		$ 145,000		$ 100,000	$ 45,000
Marketable securities (at cost; market value $60,000, 1974 and $147,000 in 1973)		55,000		150,000	(95,000)
Accounts receivable	$400,000		$425,000		
Less: Allowance for uncollectible accounts	20,000		25,000		
Net accounts receivable		380,000		400,000	(20,000)
Note receivable		10,000		8,000	2,000
Inventories (first-in first-out basis)		462,000		454,000	8,000
Total Current Assets		$1,052,000		$1,112,000	$(60,000)

Ratio Analysis

One of the fundamental expressions of a relationship between two numbers is called a *ratio*. Since accounting is concerned with the development and presentation of quantitative information, there is virtually an unlimited number of ratios which may be expressed. These expressions may take the form of percentages, decimals, or fractions, and they may be meaningful or useless. The fact that a ratio has been presented may mean only that one number has been divided by another, i.e., $200 \div 100$ is equal to a ratio of 2:1 and nothing more, unless additional information is provided.

In a later section of this chapter several common ratios used in the analyses of financial statements will be explained and illustrated.

Percentage Analysis

A rather common, and also useful, method of expressing the relationship between numbers is by *percentage calculations*. A percentage is a form of ratio calculation in which one number serves as the base and each of one or more other numbers are expressed as a percentage of that base. The calculation involves the division of the individual numbers by the base number.

The use of percentages in the analysis of financial statements is particularly helpful in the comparison of two or more periods, and it is almost essential to the evaluation of two or more companies. The use of dollar amounts in financial statements is necessary if they are to convey the financial position and results of operations of the firm; however, the dollar amounts are difficult to compare except in terms of absolute differences, e.g., sales increased $397,000. In a later section of this chapter, the percentage method of analysis will be illustrated through the use of a common size income statement.

The interpretation of financial statements may be segmented depending upon the particular interest of the decision-maker. In the remainder of this chapter, several of the more common calculations will be illustrated and discussed within the context of specific analyses. The designations given to these analyses are as follows:

Profitability analyses
Market price analyses
Working capital analyses
Special liquidity analyses
Debt-equity analyses
Common size analyses

These analyses are not the only ones available; they may not serve the needs of all users; and the titles may vary from one user to another; however, they are representative analyses and should serve as guides to any other evaluations considered necessary by the users of accounting information. The financial statements of Widget Company presented in Chapter 2, will be repeated to illustrate financial statement analyses.

The Widget Manufacturing Company, Inc.
Balance Sheet
at December 31, 1974
with Comparative Figures for 1973

ASSETS		1974		1973
Current Assets				
Cash		$ 145,000		$ 100,000
Marketable securities (at cost; market value $60,000 in 1974 and $147,000 in 1973)		55,000		150,000
Receivables:				
Accounts receivable	$400,000		$425,000	
Less: Allowance for uncollectible accounts	20,000		25,000	
Net accounts receivable		380,000		400,000
Note receivable		10,000		8,000
Inventories (first-in, first-out basis)		462,000		454,000
Total Current Assets		$1,052,000		$1,112,000
Investments				
Land held for future plant site		$ –		$ 50,000
Investment in long-term securities (at cost)		100,000		100,000
Total Investments		$ 100,000		$ 150,000
Property, Plant, and Equipment				
Land		$ 150,000		$ 100,000
Buildings	$825,000		$715,000	
Less: Accumulated depreciation	250,000	575,000	230,000	485,000
Machinery and equipment	$500,000		$450,000	
Less: Accumulated depreciation	90,000	410,000	75,000	375,000
Total Net Property, Plant, and Equipment		$1,135,000		$ 960,000

Intangible and Other Assets

	1974	1973
Patents	$ 10,000	$ 10,000
Copyright	5,000	6,000
Total Intangible and Other Assets	$ 15,000	$ 16,000
TOTAL ASSETS	$2,302,000	$2,238,000

EQUITIES	1974	1973

Current Liabilities

	1974	1973
Accounts payable	$ 260,000	$ 210,000
Notes payable (due in 1 year)	10,000	40,000
Federal income tax payable	143,000	103,000
Sales commissions payable	28,000	26,000
Total Current Liabilities	$ 441,000	$ 379,000

Long-Term Liabilities

	1974	1973
Bonds payable	$ 100,000	$ 200,000
Total Liabilities	$ 541,000	$ 579,000

Owners' Equity

	1974	1973
Common stock, $10 par value, authorized shares 60,000; issued 50,000	$ 500,000	$ 500,000
Premium on common stock	150,000	150,000
Retained earnings	1,111,000	1,009,000
Total Owners' Equity	$1,761,000	$1,659,000
TOTAL LIABILITIES AND OWNERS' EQUITY	$2,302,000	$2,238,000

The Widget Manufacturing Company, Inc.
Income Statement
for the Years 1973 and 1974

		1974		1973
Net sales		$4,797,000		$4,400,000
Cost of goods sold		3,464,000		3,163,000
Gross profit on sales		$1,333,000		$1,237,000
Operating expenses:				
Selling expense	$288,000		$275,000	
Administrative expenses	330,000		339,000	
General expenses	429,000	1,047,000	411,000	1,025,000
Operating income		$ 286,000		$ 212,000
Other income and expense Interest and dividends received		9,000		8,000
Income before income taxes		$ 295,000		$ 220,000
Federal and state income taxes		143,000		103,000
Net income		$ 152,000		$ 117,000
Earnings per share of common stock		$3.04		$2.34
Average number of shares of common stock outstanding		50,000		50,000

PROFITABILITY ANALYSES

The net income figure has little value as a single number. The amount of the net income is affected by the size of the invested capital, volume of sales, etc. These factors make any analysis of a single number extremely difficult. Thus most analyses of profitability involve a relationship between the net income and some other relevant numbers.

Profitability, as has been discussed before, is extremely important to the firm. Unless the profit earned by the firm is large enough to keep investors happy, selling future issues of stock will be difficult. This difficulty could result in a reduction of expansion plans and a curtailment of activities.

The question remains: What do investors, bankers, etc. look for as relevant measures of profitability? There are many different profitability ratios that can be developed, but those presented are considered to be the important indicators of a firm's profitability by a majority of people.

Earnings Per Share (EPS)

The earnings per share calculation is the most widely used, discussed, and abused ratio of financial analysis. EPS serves as an all-purpose, easy-to-use indicator of past profitability as a portent of the future.

The calculation of EPS was presented in Chapter 2 in a very simplified manner. The calculation was presented as:

$$\text{EPS} = \frac{\text{Net income}}{\text{Average shares of common stock}}$$

This is the basic calculation, but there are some significant points that should be evident at this point in the course. First, net income is a single number which is the result of subtracting expenses from revenues. Since there are alternative methods of computing expenses and revenues, the methods selected determine the net income. Consequently, the net income calculated is not the only net income figure that could be determined; it is the result of the choice of generally accepted accounting principles to be used in the particular situation.

Management, through the choice of alternative accounting methods, can influence the calculation of net income. For example, the choice of sum-of-years'-digits depreciation method rather than straight-line depreciation will give a higher expense and lower net income in the early years of an asset's life. Thus the EPS determined under the sum-of-years'-digits method will initially be lower than the calculation using straight-line depreciation. This will be reversed in the later years of the asset's life. There are numerous examples of other alternative methods, i.e., FIFO–LIFO, capitalize or expense cost pertaining to fixed assets, etc., and as a result, the net income is affected. The point to be noted is that net income is the result of many decisions and cannot be regarded as an infallible calculation of profit.

Because of the growing significance attached to the earnings per share calculation by investors and the desire to have such data computed on a consistent basis, the AICPA issued Opinion Number 15, *Earnings Per Share*. The opinion is a lengthy explanation of the methods to be employed in the calculation of EPS. There are many important aspects of the opinion and a full discussion is considered more appropriate for an intermediate accounting course. There are, however, certain points that should be mentioned:

1. Earnings per share data should be calculated both for earnings from continuing operations and extraordinary items. For example:

	1975	1974
Earnings per common share:		
Income before extraordinary item	$2.90	$2.55
Extraordinary item	.25	—
Net income	$3.15	$2.55

According to Accounting Principles Board Opinion Number 30, an extraordinary item is one that is both unusual in nature and infrequent in occurrence in the environment in which the business operates. Thus, given this criterion for classification as extraordinary, there are not likely to be many transactions and events classified as extraordinary. If, however, in the judgment of the company and the accountant, an item is to be classified as extraordinary, the nature of the event or transaction together with the rationale for such classification should be disclosed in a footnote.

2. For those companies whose capital structure is simple, i.e., only preferred and common stock and no potentially dilutive convertible securities, options, rights, or warrants, the weighted average number of shares outstanding can be used in the calculation. Potentially dilutive refers to those convertible securities, options, rights, or warrants that, upon conversion or exercise, could dilute earnings per common share. Therefore, the basic calculation of EPS for a simple capital structure is:

$$\frac{EPS-}{ordinary} = \frac{Net\ income - preferred\ dividends}{Weighted\ average\ of\ common\ shares\ outstanding}$$

$$\frac{EPS-}{extraordinary} = \frac{Extraordinary\ gain\ or\ (loss)}{Weighted\ average\ of\ common\ shares\ outstanding}$$

The weighted average number of shares is basically the same calculation as the weighted average method of inventory valuation. In this case, the shares outstanding are multiplied by the number of months they were outstanding and divided by the number of months in a year.

3. Corporations with capital structures that are not simple, e.g., with outstanding convertible securities, must present two types of

earnings per share data. The first calculation, called *primary earnings per share*, is based on the outstanding common stock plus those securities that are *in substance* equivalent to common shares. These securities are called *common stock equivalents*. There are many criteria for common stock equivalents, but these are considered to be too technical for an introductory course.

In addition to primary earnings per share, a second calculation is made using the common stock equivalents and all potentially convertible securities that could dilute earnings per share. This calculation is referred to as *fully diluted earnings per share*.

An illustration of the EPS data for both primary and fully diluted earnings per share is presented as follows:

	1975	1974
Primary earnings per share:		
Income before extraordinary items	$2.90	$2.55
Extraordinary item	.25	—
Net income	$3.15	$2.55
Fully diluted earnings per share:		
Income before extraordinary items	$2.82	$2.50
Extraordinary items	.22	—
Net income — fully diluted basis	$3.04	$2.50

The complexity of the earnings per share calculation should be evident. The average investor tends to make a simplistic and cursory review to determine whether EPS have increased or decreased. The ultimate investment decision may be correct, but a casual review of EPS is generally unwise considering the intricacies of the actual calculation.

Net Income to Net Sales

In the calculation of the ratio of net income to net sales, the net income after taxes, but before extraordinary items, is used so that a percentage of the sales dollar available for reinvestment or dividend distribution can be determined. The actual calculation is:

$$\text{Percentage return on each sales dollar} = \frac{\text{Net income after taxes}}{\text{Net sales}}$$

This percentage is important as a measure of the return available to stockholders on each dollar of sales. Naturally, a portion of the income will be reinvested, but this ratio reflects the percentage of each sales dollar that is available for distribution.

In the example presented, the ratio of net income to net sales is:

$$\begin{array}{cc} 1974 & 1973 \\ \dfrac{\text{Net income}}{\text{Net sales}} = \dfrac{\$152,000}{\$4,797,000} = 3.2\% & \dfrac{\$117,000}{\$4,400,000} = 2.7\% \end{array}$$

The increase in net income to net sales is reflected in the increase in the earnings per share. Note that the net income increased $35,000 with an increase of $397,000 in net sales. Thus the increase in the marginal income was $35,000/$397,000 = 9\%. Clearly, an expansion of sales with an increase in marginal income of 9% is in the best interest of the firm. The increase may have been the result of certain economies of scale that were achieved. Management should investigate the reason for the increase in order to make appropriate decisions concerning further expansion.

Net Income to Owners' Equity

The ratio of net income to owners' equity reflects the return on owners' equity that the firm was able to earn. If the return earned by the firm is insufficient to attract or maintain investors, the firm will have difficulty expanding operations because additional capital will not be available.

In the example, the rate of return is:

$$\begin{array}{cc} 1974 & 1973 \\ \dfrac{\text{Net income}}{\text{Owners' equity}} = \dfrac{\$152,000}{\$1,761,000} = 8.6\% & \dfrac{\$117,000}{\$1,659,000} = 7\% \end{array}$$

Investors should compare these rates of return with returns on other types of investments.

Net Income to Total Assets

As indicated in Chapter 2, readers of financial statements may be interested in the rate of return the firm is earning on the assets used in the business. This information can be obtained by relating net income to the total assets. In the example, the rate of return on total assets is:

$$\begin{array}{ccc} & \underline{1974} & \underline{1973} \\ \dfrac{\text{Net income}}{\text{Total assets}} = & \dfrac{\$152,000}{\$2,302,000} = 6.6\% & \dfrac{\$117,000}{\$2,238,000} = 5.2\% \end{array}$$

This information can be used by management in evaluating the present return on invested assets as a guide for returns expected on future purchases of assets. Since this ratio is after taxes and does not consider the time value of money, adjustments should be made for these factors.

There are other ratios indicating various aspects of profitability that could be developed. Actually, any ratio relating net income to another number is a profitability indicator. The ratios presented are those that are generally used in evaluating the profitability of a particular firm. In terms of importance and/or use, earnings per share is given the most weight. As a handy reference, EPS is a good indicator of profitability but, in a particular situation, another ratio or ratios might be more informative.

MARKET PRICE ANALYSES

There are two common ratios used to evaluate the market price of common stock. The *price-earnings* ratio (P-E) indicates the multiple of earnings which investors are willing to pay for the stock. The *book value per share* is the total common stockholders' capital, i.e., total stockholder equity less amount to be paid preferred stockholders at liquidation, divided by the number of common shares outstanding. While book value per share may have little relation to market value, it is often used as the starting point in arriving at market price in the purchase or sale of a business.

Price-Earnings Ratio (P-E)

The price-earnings ratio is calculated by relating the market price of the stock to the earnings per share. If the market price of Widget Company stock was $50 in 1974, the P-E ratio would be 16.5:1. The calculation is:

$$\text{Price-earnings ratio} = \frac{\text{Market price}}{\text{Earnings per share}} = \frac{\$50}{\$3.04} = 16.5:1$$

The P-E ratio means that, at the present time, the stock is 16.5 times the earnings per share. If the relationship between EPS and market price remains constant, the P-E ratio can be used as an indicator of future market prices. For example, if earnings for 1975 were projected to be $4.00 per share, the expected market price would be $66. Therefore, as an investor, you should buy this particular stock at $50 with the expectation of the price increasing to $66. Naturally, if earnings fail to reach $4 or, even worse, fall from $3.04, the price of the stock would be 16.5 times whatever earnings are achieved.

The question may be asked: Why does this firm have a P-E ratio of 16.5? Unfortunately, the answer is more difficult than the question. The ratio is really a composite of investor expectations concerning the nature of the industry, rules-of-thumb developed within the financial sector, and industry-wide projections. Consequently, no one P-E ratio affixed to a firm remains constant. The ratio changes daily as investor expectations concerning future earnings change; however, the ratio will not change dramatically unless there is a significant change in the expectations concerning the firm.

A firm may have increased earnings and at the same time a decreased market price. The reason for this is lower investor expectations concerning the firm and/or future industry profit potential. Since investor expectations play a significant role in the P-E ratio, the firm should attempt to demonstrate growth potential if a high P-E ratio is desired.

Book Value per Share

Book value per share is computed by dividing the common stockholder net worth by the number of shares of outstanding common stock. The calculation is illustrated as:

$$\frac{\text{Book value}}{\text{per share}} = \frac{\text{Assets} - \text{Liabilities and preferred stock claims}}{\text{Number of shares of common stock outstanding}}$$

The common stockholders' net worth is the value obtained by deducting the book value of the liabilities and preferred stock claims from the book value of the assets. Consequently, market values are ignored. Since the purpose of this calculation is to obtain the book value of a share of common stock, the par value or liquidating value of preferred stock is subtracted in determining the net worth available to the common stockholders.

In the example, the book value per share of common stock is:

	1974	1973
Book value per share	$\dfrac{\$1,761,000}{50,000} = \$35.22$	$\dfrac{\$1,659,000}{50,000} = \$33.18$

The relevance of this ratio can be questioned since market values are ignored. In truth, unless a company is nearing liquidation or the book and market values of assets are equal, this ratio means little. Perhaps the best thing that can be said is that this should be the floor for any market price of the stock. A market price lower than the book value might indicate that the firm is worth more in liquidation than as a going concern.

WORKING CAPITAL ANALYSES

Working capital is defined as the excess of the current assets over the current liabilities. The amount of working capital is important because of the liquidity necessary to operate the business on a day-to-day basis. The amount of working capital as an absolute number is not as important as the relationship between the current assets and current liabilities. For example, working capital of $200 is not as impressive as working capital of $50,000; however, if the $200 is the difference between assets of $300 and liabilities of $100 and the $50,000 is the difference between $1,050,000 of assets and $1,000,000 of liabilities, then the absolute difference is misleading. The relationship between current assets and current liabilities is commonly expressed in two ratios.

Current Ratio

The current ratio is calculated by dividing the current assets by the current liabilities. The current ratio for Widget Company is:

$$1974$$

$$\text{Current ratio} = \frac{\text{Current assets}}{\text{Current liabilities}} = \frac{\$1,052,000}{\$441,000} = 2.39{:}1$$

and:

$$1973$$

$$\frac{\$1,112,000}{\$379,000} = 2.93{:}1$$

This ratio means that for every dollar of liabilities there is a certain dollar amount of assets available for their payment. An increase in the ratio indicates a more liquid position because there is a greater amount of assets available for payment of the liabilities. A decrease in the ratio, as in the example, indicates a possible impairment of short-term liquidity.

The advantage of using the current ratio rather than the amount of working capital is that ratios can be related and compared while it is difficult to compare absolute numbers. In addition, the current ratio can be used for comparison with industry-wide figures or previous years. In this manner, a yardstick is developed against which the firm can measure the adequacy of its liquidity position.

Acid-Test Ratio (Quick Ratio)

For some firms, the current ratio is not the best measure of liquidity. The time lag between the purchase of inventory and the sale and ultimate receipt of cash may be such that the current ratio does not really reflect the liquidity position. In these situations, the *acid-test ratio* is used as the indicator of liquidity.

This ratio is the *relationship of cash, marketable securities, accounts receivable, and short-term notes receivable to the current liabilities*. In the example, the acid-test ratio is calculated as:

$$1974$$

$$\frac{\text{Acid-test}}{\text{ratio}} = \frac{\$145,000 + \$55,000 + \$380,000 + \$10,000}{\$441,000} = 1.34{:}1$$

and:

$$1973$$

$$\frac{\$100,000 + \$150,000 + \$400,000 + \$8,000}{\$379,000} = 1.74{:}1$$

This ratio tests the ability of the firm to meet sudden demands upon its liquidity. In the example, as of December 31, 1974, there are $1.34 of liquid assets available to pay each dollar of current liabilities. This amount would seem to be adequate, but a continued decrease of $0.40 per year ($1.74 − $1.34) cannot be tolerated.

Either the current or the acid-test ratios can be used to measure the liquidity position of a firm. Care must be used in evaluating these ratios because of possible window dressing. This means that events may take place at the end of the year to make the ratios look better. For example, the payment of an accounts payable at the end of the year will increase the current ratio if the ratio is greater than 1. If the ratio is less than 1, the same transaction will decrease the ratio. For example, assume current assets of $100 and current liabilities of $50:

$$\text{Current ratio} = \frac{\text{Current assets}}{\text{Current liabilities}} = \frac{\$100}{\$50} = 2{:}1$$

A payment of current liabilities in the amount of $25 increases the ratio:

$$\text{Current ratio} = \frac{\text{Current assets}}{\text{Current liabilities}} = \frac{\$75}{\$25} = 3{:}1$$

In addition to the payment of a liability, other events may affect the ratio. Note that the creation of a liability will decrease the ratio if it is greater than 1. Other aspects of possible window dressing include sales of fixed assets, sales of bonds or stock, and the conversion of short-term debt into long-term debt.

SPECIAL LIQUIDITY ANALYSES

The importance of liquidity to the successful operation of a business was emphasized in Chapter 5 and also as a part of the working capital analyses in the preceding section of this chapter. In manufacturing and retailing firms, inventory and accounts receivable comprise a substantial portion of the current assets. Consequently, these two accounts have a major influence on the liquidity of such firms and give rise to the need for special analyses.

Inventory Turnover

Any manufacturing or retailing firm must maintain a quantity of inventory in order to continue as a going concern; however, an excess quantity of inventory means additional costs of holding the inventory. In the Appendix to Chapter 7, the Economic Order Quantity model was described as an attempt to minimize the cost of ordering and holding inventory. The model, as presented, did not include any opportunity cost or charges for reduced liquidity. In this context, the term *opportunity cost* refers to the alternative uses for the resources invested in inventory.

The inventory turnover calculation does not determine the opportunity costs, but it does draw attention to a potential under- or overstocked situation. The calculation of inventory turnover is:

$$\text{Inventory turnover} = \frac{\text{Cost of goods sold}}{\text{Average inventory}}$$

This ratio simply indicates the liquidity of the inventory, i.e., the number of times, on the average, that the inventory turned over or was sold during the period.

In this calculation, the cost of goods sold is taken directly from the income statement. The average inventory is commonly derived by taking the average of the beginning and ending balances shown on the comparative balance sheets. A more precise average may be determined if monthly inventory balances are available; however, this information is not usually presented in external statements.

The inventory turnover of Widget Company is:

$$\frac{\text{Inventory}}{\text{turnover}} = \frac{\text{Cost of goods sold}}{\text{Average inventory}} = \frac{\$3,464,000}{\$458,000} = 7.56 \text{ turnovers per year}$$

The number of turnovers per year must be viewed as an average of the inventory and not as an indication of the turnover of individual items. In addition, the turnover rate for Widget Company cannot be judged as good or bad except in relation to past years and to similar firms. The operating cycle, i.e., the length of time necessary to go from a cash position, to inventory, to receivables, to cash again, varies between industries. Therefore, the number of turnovers in a grocery store may greatly exceed the turnovers in a hardware store.

Once the inventory turnover has been calculated, it may be useful to determine the number of days supply of inventory on hand. The calculation is:

$$\frac{\text{Number of}}{\text{days supply}} = \frac{365 \text{ Days}}{\text{Inventory turnover}} = \frac{365}{7.56} = 48 \text{ days supply}$$

This information can then be compared to previous years and other firms. Also, the number of days supply must be related to the time necessary to acquire inventory.

Accounts Receivable Turnover

Once the inventory is sold, the company has achieved a major step toward liquidity; however, in a credit society such as that in the United States, a substantial portion of sales is made on credit. Insofar as liquidity is concerned, this means that while the company is a step closer to cash there is still a waiting period. The importance of sound credit policies and collection procedures for accounts receivables was discussed in Chapter 6.

The purpose of the accounts receivable analysis is to determine the speed with which credit sales are collected. A slow-down in collection may indicate a poor credit policy or ineffectiveness of collection procedures.

The calculation for accounts receivable turnover is:

$$\frac{\text{Accounts receivable}}{\text{turnover}} = \frac{\text{Net credit sales}}{\text{Average accounts receivable (net)}}$$

Since this ratio requires information relative to credit sales, the external user of financial statements may not be able to make the analysis. Most published statements do not include a breakdown between cash and credit sales, in which case the company must provide the ratio or, as an approximation, total sales may be used. Obviously, this will not be completely valid, but it does provide for comparison with previous years and other firms in the industry assuming that the ratio of cash to credit sales is approximately the same from year to year and firm to firm.

As in the case of inventory, the average accounts receivable may be either average monthly balances or average yearly balances. The ending balance may also be used if the previous period balance is not available. Also, note that the calculation uses net accounts receivable, i.e., accounts receivable less allowance for uncollectible accounts.

The accounts receivable turnover for Widget Company will be illustrated by assuming that all sales were credit sales:

$$\text{Accounts receivable turnover} = \frac{\text{Net credit sales}}{\text{Average accounts receivable (net)}} = \frac{\$4,797,000}{\$390,000} = 12 \text{ turnovers per year}$$

The number of turnovers can then be used to determine the average number of days between the sale and the collection of accounts receivable.

$$\text{Day's accounts receivable outstanding} = \frac{365 \text{ days}}{\text{Receivable turnover}} = \frac{365}{12} = 30 \text{ days}$$

This means that if the company has a discount policy, for example 2/10, n/30, any collections made early are offset by collections which exceed the time limit. Whatever the average number of days it takes to collect the receivables, the collection period should be related to the credit policy of the firm.

DEBT TO EQUITY ANALYSES

The characteristics which distinguish debt from equity were discussed in Chapter 10. As indicated at that point, there are certain securities, such as convertible bonds, which have characteristics of both debt and equity. With that qualification in mind, plus an awareness of the various components of the Liabilities and Owners' Equity sections of the balance sheet, an evaluation can be made of the relationship between debt and equity.

A major factor in the debt-equity determination revolves around risk. The return on investment is normally related to risk, i.e., the greater the risk, the greater the return. In the case of owners, the rate of return may be zero or negative, but the potential return is much higher than for creditors who have a fixed rate of return along with certain assurances of payment.

In the case of a corporation, that assurance is provided, first of all by the expectation that the firm will earn a profit, and secondly by the amount of investment which the owners have in the firm. If the owners have only nominal investment while creditors have provided most of the resources, then the creditors have little assurance of payment other than through profits. Consequently, the calculation of a debt to equity ratio provides meaningful information for evaluating the relative position of creditors and owners. Such a ratio is also useful for comparing one firm to another where dollar amounts might be misleading or difficult to relate.

The debt to equity ratios for Widget Company as of December 31, 1973 and 1974 are:

$$\underline{1974}$$

$$\text{Debt-equity ratio} = \frac{\text{Liabilities}}{\text{Owners' equity}} = \frac{\$541,000}{\$1,761,000} = 0.31{:}1$$

and:

$$\underline{1973}$$

$$\frac{\$579,000}{\$1,659,000} = 0.35{:}1$$

This means that the owners of Widget Company, as of December 31, 1974, have invested $1 for each $0.31 that the creditors have loaned the firm. This is a slight improvement from the creditors' point of view over the previous year when the owners had invested $1 for each $0.35 loaned the firm by the creditors.

COMMON SIZE ANALYSES

As discussed in a previous section, percentage analysis is a particularly good method for comparing two or more years of data. To illustrate, a common size income statement is presented with net sales serving as the base:

Widget Company
Common Size Income Statement

	1974	1973
Net sales	100.0%	100.0%
Cost of goods sold	72.2%	71.9%
Gross profit on sales	27.8%	28.1%
Operating Expenses:		
Selling expenses 6.0%		6.3%
Administrative expenses 6.9%		7.7%
General expenses 8.9%		9.3%
Total operating expenses	21.8%	23.3%
Operating income	6.0%	4.8%
Other income	0.2%	0.2%
Income before income tax	6.2%	5.0%
Federal and state income tax	3.0%	2.3%
Net income	3.2%	2.7%

The changes that affected net income are easier to isolate through this type of analysis. There was a small increase in cost of goods sold, but this was more than offset by the decreases in each of the operating expenses. The income tax, naturally, increased because of the higher income before taxes. Thus comparisons of two or more years can be made more readily if percentages are developed rather than using dollar amounts.

FOOTNOTES

A substantial amount of information about the financial position of the firm is frequently presented in footnotes to the financial statements. When footnotes are presented they are considered to be an integral part of the financial statements. As such, the user of financial statements must consider the footnotes when attempting to analyze the firm.

All financial statements, certified by a CPA, must contain information relative to the significant accounting policies employed by the firm. Among the policies that must be disclosed are those related to:

1. Depreciation methods
2. Amortization of intangibles
3. Inventory pricing

There are frequently footnotes which disclose information other than accounting policies. For example, information may be presented about pension plans, lease commitments, convertible securities, etc. Thus there is potentially an enormous amount of information in the footnotes to financial statements such that any analysis of a firm should include a review of the footnotes.

SUMMARY

There are several points about financial statement analysis which should be remembered:

1. Any relationship between two numbers can be called a ratio. The fact that a ratio has been calculated does not necessarily make it useful for decision purposes.
2. Ratios, like other facts, are subject to misrepresentation.
3. The analysis employed should be selected on the basis of the users' particular needs.
4. The dollar amounts in financial statements are the result of certain choices as to the accounting methods to be used; therefore, the ratios are no more precise than the statements.

QUESTIONS

13-1. Why is the analysis of financial statements most appropriately thought of as the job of the financial statement user?

13-2. What is meant by comparative financial statements? Why are they useful in analyzing a company's operations?

13-3. The analysis of financial statements is performed by a number of different groups. Indicate some of the reasons the following groups would have for analyzing financial statements:
   (a) Managers
   (b) Stockholders
   (c) Prospective investors
   (d) Government organizations
   (e) Labor unions

13-4. Is it possible to have different earnings per share and still use generally accepted accounting principles?

13-5. Differentiate between the computation of earnings per share under a simple capital structure and a complex capital structure.

13-6. Of what significance are the following ratios in analyzing a company's operations?
(a) Net income to net sales
(b) Net income to owners' equity
(c) Net income to total assets

13-7. Of what significance is the fact that the book value of a company's outstanding stock is greater than the market value?

13-8. Explain the calculation of the following two ratios and discuss their significance:
(a) Price-earnings ratio
(b) Book value per share

13-9. Define working capital. Explain why the absolute amount of working capital of a particular company may not provide adequate information about the current financial condition of the firm.

13-10. Explain the calculation of the following ratios and discuss their significance:
(a) Current ratio   (c) Inventory turnover
(b) Acid-test ratio   (d) Accounts receivable turnover

13-11. Differentiate between owners and creditors of a corporation in terms of the risks each assumes relative to his expected return on investment.

13-12. What is common size analysis? What is the advantage of this type of analysis over the presentation of actual dollar amounts?

PROBLEMS

Procedural Problems

13-1. Ross Corporation has the following trial balance:

	Dr.	Cr.
Cash	$ 4,000	
Marketable securities	3,000	
Accounts receivable	8,000	
Inventory (no change since last year)	15,000	
Building (net of depreciation)	30,000	
Accounts payable		$ 8,000
Salaries payable		2,000
Notes payable (due in 10 years)		10,000
Common stock ($25 par)		25,000
Sales (all credit)		85,000
Cost of goods sold	57,000	
Other expenses	13,000	
	$130,000	$130,000

REQUIRED. Using the foregoing trial balance, compute the following:
- (a) Current ratio
- (b) Acid-test ratio
- (c) Inventory turnover
- (d) Accounts receivable turnover
- (e) Average number of days to collect accounts receivable

13-2. The following balance sheet was prepared from the records of Goebel Corporation as of December 31, 1974:

Goebel Corporation
Balance Sheet
as of December 31, 1974

Cash	$ 25,000	Accounts payable	$ 45,000
Marketable		Taxes payable	5,000
securities	20,000	Mortgage payable	25,000
Accounts		Common stock	
receivable	40,000	($50 par)	100,000
Merchandise		Retained earnings	50,000
inventory	60,000		
Fixed assets			
(net)	80,000		
	$225,000		$225,000

Net income for the year amounted to $16,000.

REQUIRED. Compute the following ratios:
- (a) Current ratio
- (b) Acid-test ratio
- (c) Earnings per share
- (d) Net income to owner's equity
- (e) Book value per share

Based on your analysis, what will be the price of the common stock?

13-3. The A. J. Smith Company has working capital of $100,000, a current ratio of 3:1, and an acid-test ratio of 1.5:1. For each of the following transactions, indicate the effect on the company's working capital, current ratio, and acid-test ratio. Indicate the effect in terms of increase, decrease, or no change. Assume each transaction is independent, i.e., not cumulative:
- (a) Collected a $2,000 account receivable.
- (b) Paid a $5,000 account payable.
- (c) Sold inventory costing $10,000 for $12,000 on account.
- (d) Declared a $10,000 cash dividend.
- (e) Purchased a building giving a mortgage due in ten years.
- (f) Paid the dividend declared in (d).
- (g) Purchased merchandise inventory on account.
- (h) Borrowed $5,000 giving a 60-day, 6% note.
- (i) Sold marketable securities costing $8,000 for $6,000 cash.
- (j) Accrued $2,500 of salaries expense.

13-4. Selected items from the financial statements of Longfield and Barker Corporations are presented as follows:

	Longfield	Barker
Current assets	$ 40,000	$ 75,000
Fixed assets (net)	80,000	220,000
Total liabilities	25,000	60,000
Retained earnings[a]	45,000	135,000
Common stock	$50 par	$50 par
Net sales	150,000	400,000
Net income	25,000	50,000

[a]Includes current net income.

REQUIRED. Perform a complete profit analysis on the preceding two companies. Indicate which company has the stronger profit picture.

13-5. The following financial statements were taken from the books of Baker Company:

Baker Company
Balance Sheet
as of December 31, 1974

ASSETS

Current Assets

Cash		$ 20,000	
Marketable securities		14,000	
Accounts receivable	$ 35,000		
Less: Allowance for uncollectible accounts	2,000	33,000	
Merchandise inventory		47,000	
Prepaid rent		6,000	
Total Current Assets			$120,000

Fixed Assets

Land		$ 66,000	
Building	$100,000		
Less: Accumulated depreciation	6,000	94,000	
Equipment	$ 48,000		
Less: Accumulated depreciation	3,000	45,000	
Total Fixed Assets			205,000
TOTAL ASSETS			$325,000

EQUITIES

Current Liabilities

Accounts payable	$ 40,000	
Accrued wages payable	6,000	
Taxes payable	4,000	
Total Current Liabilities		$ 50,000

Long-Term Liabilities

Notes payable (due in 1980)	$ 20,000	
Mortgage payable (due in 1985)	40,000	
Total Long-Term Liabilities		60,000
Total Liabilities		$110,000

Owners' Equity

Common stock $25 par value	$100,000	
Retained earnings	115,000	
Total Owners' Equity		215,000
TOTAL LIABILITIES AND OWNERS' EQUITY		$325,000

Baker Company
Income Statement
for the Year Ended December 31, 1974

Gross sales		$545,000
Less: Sales returns		5,000
Net sales		$540,000
Cost of goods sold:		
Beginning inventory	$ 68,000	
Plus: Purchases	248,000	
Goods available for sale	316,000	
Less: Ending inventory	47,000	
Cost of goods sold		269,000
Gross profit on sales		$271,000
Operating expenses:		
Selling expenses	$ 49,200	
Administrative expenses	71,000	
General expenses	90,600	
Total operating expenses		210,800

Operating income	**$ 60,200**
Income tax expense	**$ 30,100**
Net income	**$ 30,100**

REQUIRED. Perform the following analyses on the preceding financial statements:
(a) Profitability
(b) Market price (current market price is $90 per share)
(c) Working capital
(d) Special liquidity
(e) Debt to equity
For each of the preceding analyses indicate the significance of the results obtained. Consider this section in terms of the advisability of investing in the stock of Baker Company.

13-6. The following information is available concerning the financial data of two separate companies:

	Company A	Company B
Net income after taxes	$ 25,000	$ 31,500
Preferred stock outstanding	1,000	5,000
Preferred par value	$ 100	$ 50
Dividend rate on preferred	5%	4%
Common shares outstanding	2,000	3,000
Par value of common	$ 50	$ 50
Retained earnings	$ 80,000	$ 50,000
Net sales	$120,000	$100,000

Assume the dividends on preferred stock are cumulative and have been paid in all prior years, but have not been paid for the current year. Retained earnings as shown previously include the current year's net income. Dividends on preferred stock should be distributed prior to performing the calculations required.

REQUIRED. Calculate the following:
(a) Earnings per common share
(b) Net income to net sales
(c) Net income to owners' equity
(d) Book value per share

13-7. Following are the income statements of Sheffield Company for the years 1974 and 1975:

	1974		1975	
Gross sales		$123,600		$153,000
Sales returns		3,600		3,000
Net sales		$120,000		$150,000
Cost of goods sold		82,440		105,600
Gross profit		$ 37,560		$ 44,400
Operating expenses:				
Selling expenses	$8,640		$11,250	
General expenses	7,560		10,950	
Administrative expenses	6,600	22,800	9,450	31,650
Operating income		$ 14,760		$ 12,750
Dividend and interest income		4,800		6,000
Income before tax		$ 19,560		$ 18,750
Federal and state taxes		9,720		9,000
Net income		$ 9,840		$ 9,750

REQUIRED
(a) Prepare common size income statements for the years 1974 and 1975.
(b) Point out through common size analysis the major factors contributing to the stability of net income in a period of rising sales.

Conceptual Problems

13-8. Construct a balance sheet based upon the following ratios with the following classifications:
(a) Current asset (excluding merchandise inventory)
(b) Merchandise inventory
(c) Total fixed assets
(d) Current liabilities
(e) Long-term liabilities
(f) Total equity (includes only common stock and retained earnings)
(g) Retained earnings (includes current net income)

Current ratio	= 3:1
Acid-test ratio	= 1.3:1
Net income to total assets	= 12%

Debt to equity   = .4:1

Net income to owners' equity   = 16.8%

Common Stock is twice as large as retained earnings. Current liabilities are equal to long-term liabilities. Net income is equal to $12,600.

13-9. The president of Weston Company is concerned about the condition of his company's fixed assets. The assets were purchased when the company began business ten years ago. The accumulated depreciation on a major portion of the assets is 90% of their cost. Because of this situation, the president has asked the chief accountant to determine a solution to this problem. The company needs a loan of $100,000 to purchase new fixed assets in order to continue operations. Based upon the financial data presented, you are to determine whether or not you would extend the Weston Company the loan they desire. Calculate pertinent ratios in defending your answer:

Balance Sheet Information

	12/31/74	12/31/75
Cash	$ 8,000	$ 5,000
Accounts receivable (net)	38,000	46,000
Inventory	102,000	96,000
Accounts payable	52,000	61,000
Salaries payable	8,000	7,000
Long-term notes payable (6% interest, due in 1978)	20,000	20,000
Common stock ($50 par)	200,000	200,000
Retained earnings	12,000	15,000

Income Statement Information

Sales (all credit)		$105,000
Cost of goods sold		33,000
Gross profit		$ 72,000
Operating expenses:		
Selling expenses	$ 18,000	
Administrative expenses	28,500	
General expenses	20,000	66,500
Operating income		$ 5,500
Federal and state taxes		2,500
Net income		$ 3,000

13-10. Following are certain ratios computed from the financial statements of Oben Company and Berger Company. Both companies are in the same industry:

	Oben	Berger
Current ratio	5:1	2:1
Acid-test ratio	0.5:1	1:1
Net income to sales	10%	13%
Inventory turnover	4	10
Receivables turnover	6	12
Debt to equity	0.6:1	0.3:1
Net income to total assets	4.3%	9.1%
Book value per share	$51	$18.50

REQUIRED. Analyze the financial condition of both companies as a possible investment.

13-11. The Theresa Corporation showed the following account balances relative to its income for the past five years:

	1971	1972	1973	1974	1975
Sales	$800,000	$840,000	$780,000	$750,000	$810,000
Net income	33,600	37,000	32,500	30,000	36,450
Total assets	390,000	410,000	420,000	435,000	440,000
Owners' equity	180,000	225,000	240,000	230,000	235,000
Shares outstanding	6,000	7,000	7,000	6,500	6,500
Market value per share	56	54	50	52	58

REQUIRED
(a) Perform a complete profitability and market price analysis on the accounts of Theresa Corporation.
(b) Indicate any favorable or unfavorable trends which you observe from your analysis of this information.
(c) Would you recommend this stock as a potential investment when compared with stocks presently listed in *The Wall Street Journal?*

13-12. The following data represent the financial information as published in the annual reports of the Haywood and Irwin Corporations, respectively:

	Haywood	Irwin
Cash	$ 13,000	$ 22,000
Accounts receivable (net)	17,500	34,600
Marketable securities	10,000	9,000
Inventory	26,500	51,200
Prepaid expenses	3,000	9,000
Land	65,000	90,000

	Haywood	Irwin
Other tangible fixed assets	$110,000	$215,000
Accumulated depreciation	25,000	48,000
Long-term investments	15,000	30,000
Accounts payable	10,300	22,700
Accrued salaries	6,000	8,750
Notes payable — short-term	17,700	46,300
Taxes payable	1,000	6,115
Long-term liabilities	45,000	87,500
Common stock ($50 par value)	100,000	200,000
Capital in excess of par value	20,000	15,000
Retained earnings	35,000	26,435
Sales (90% credit)	105,000	238,500
Cost of goods sold	51,600	121,600
Administrative expenses	14,700	28,900
Operating expenses	26,000	51,100
Income taxes	2,200	10,500
Current market price per share	$57.75	$118.80

REQUIRED
(a) The chapter is divided into sections which describe the various kinds of ratios used in financial statement analysis. You are to select the one ratio from each of the five sections which you feel is the most relevant from that section. Compute those ratios for the two companies presented in the problem. Briefly indicate your reasons for the ratios you selected. (Carry computations to two decimal places.)
(b) Based upon your analysis, which stock is the best buy?

14

SOURCES OF ACCOUNTING PRINCIPLES AND CONTEMPORARY ISSUES

EDUCATIONAL OBJECTIVES

The material in this chapter is designed to achieve several educational objectives. These include:

1. A knowledge of the sources of accounting principles and standards and the organizations that influenced them, especially the Accounting Principles Board and the Financial Accounting Standards Board.

2. An awareness that, except for governmental agencies, there is no legal authority for the establishment of accounting standards; thus the profession must assume this responsibility, and it has in the form of the American Institute of Certified Public Accountants.

3. An insight into some of the current problems that exist in accounting, e.g., materiality, historical cost and current value, matching, and what constitutes relevant information.

4. A basic understanding of the purpose and structure of the Financial Accounting Standards Board.

The preceding chapters have explained the processes involved in identifying, measuring, and communicating accounting information for the benefit of the users of financial statements. In each of these processes there may be alternative methods. For example, depreciation may be recorded using straight-line depreciation or double-declining balance depreciation; some assets are measured at current value or at historical cost, e.g., use of lower cost or market; and some information can be shown in the financial statements or in the footnotes. Thus the fact that alternatives exist means that there must be a decision concerning the alternative to be used.

For accounting purposes, the underlying criterion for the selection of the alternative to be used in any situation is that it must be in conformity with generally accepted accounting principles. Throughout the text, reference was made to generally accepted accounting principles such as relevance, consistency, matching, and objectivity. These and other generally accepted accounting principles provide guidance to the accountant in the exercise of judgment in the selection of alternatives.

The purpose of this chapter is to present a discussion of three aspects of generally accepted accounting principles:

1. The meaning of the expression.
2. The major sources of the principles.
3. Some contemporary problems in the application of these principles.

THE MEANING OF "GENERALLY ACCEPTED ACCOUNTING PRINCIPLES" (GAAP)

The committee on terminology of the American Institute of Certified Public Accountants (AICPA) indicated the scope of the accountants' work in the following manner:

> It is desirable that the accountant conceive of his work as a complex problem to be solved and of his statements as creative works of art, and that he reserve to himself the freedom to do his work with the canons of the art constantly in mind and as his skill, knowledge, and experience best enable him. Every art must work according to a body of applicable rules, but it also must reserve the right to depart from the rules whenever it can thereby achieve a better result.[1]

As the previous paragraph indicates, the accountant is free to do what he believes is best in a particular situation; however, this freedom could lead to every accountant developing ideas and concepts that only he would understand. In order to avoid this situation, there must be certain basic guidelines applicable to accounting which can be understood by any accountant. Thus accountants need principles. The definition of a principle as used in accounting means "a general law or rule adopted or professed as a guide to action; a settled ground or basis of conduct or practice."

According to the AICPA, this definition:

> . . . comes nearest to describing what most accountants, especially practicing public accountants, mean by the word *principle*. Initially, accounting postulates are derived from experience and reason; after postulates so derived have proved useful, they become accepted as principles of accounting. When this acceptance is sufficiently widespread, they become a part of the "generally accepted accounting principles" which constitute for accountants the canons of their art.
>
> Care should be taken to make it clear that, as applied to accounting practice, the word *principle* does not connote a rule from which there can be no deviation. An accounting principle is not a principle in the sense that it admits of no conflict with other principles. In many cases the question is which of several partially relevant principles has determining applicability.[2]

In order to meet the responsibility placed upon him by the preceding definition, the accountant must possess adequate knowledge of the subject and have the ability to exercise good judgment. He must be able to determine whether the accounting practices used by a business for external reporting are generally accepted.

The preceding explanation of generally accepted accounting principles is commonly stated more precisely among accountants. The Council of the AICPA has asserted that generally accepted accounting principles are "those principles of accounting which have substantial authoritative support."[3] The Council went on to say that statements by the Accounting Principles Board on accounting matters constituted "substantial authoritative support," but there are also other sources of support. Among the sources which provide substantial authoritative support for accounting principles are:

1. Pronouncements issued by the Financial Accounting Standards Board and its predecessors, the Accounting Principles Board and the Committee on Accounting Procedure.
2. Pronouncements by committees of the American Accounting Association.
3. Regulatory pronouncements issued by the Securities and Exchange Commission.
4. Pronouncements by regulatory agencies concerning the accounting to be employed in reporting to the agency.
5. Practices commonly found in business.
6. Writings of individual accountants which appear in the professional journals.

In this section, special emphasis will be given to the role of various organizations in providing authoritative support. This does not imply that individuals and business practices should be ignored; however, because they are so numerous, they cannot be discussed effectively in an introductory manner.

American Institute of Certified Public Accountants

As discussed in Chapter 1, the AICPA is an organization composed of accountants who have qualified as certified public accountants. Since the majority of the membership is composed of public accountants, it is only natural that pronouncements of the various AICPA committees have had the greatest influence on the practice of accountancy.

The influence on practice has emanated from the reports of various special committees, *Accounting Terminology Bulletins* issued by the Committee on Terminology, *Research Studies* of the Accounting Research Division, *Accounting Research Bulletins* issued by the Committee on Accounting Procedure, *Opinions* and *Statements* of the Accounting Principles Board, and pronouncements of the Financial Accounting Standards Board. The Committee on Accounting Procedures was organized in 1939 and continued until replaced by the Accounting Principles Board in 1959. The APB was the most influential and authoritative unit within the AICPA from 1959 until 1973 when it was replaced by the Financial Accounting Standards Board (FASB).

The Committee on Accounting Procedure explained the applicability and authority of its opinions (the APB and the FASB have continued to operate accordingly) in the following paragraphs:

> Underlying all committee opinions is the fact that the accounts of a company are primarily the responsibility of management. The responsibility of the auditor is to express his opinion concerning the financial statements and to state clearly such explanations, amplifications, disagreement, or disapproval as he deems appropriate.
>
> The principal objective of the committee has been to narrow areas of difference and inconsistency in accounting for determining the suitability of accounting practices reflected in financial statements and representations of commercial and industrial companies. Accordingly, except where there is a specific statement of a different intent by the committee, its opinions and recommendations are directed primarily to business enterprises organized for profit.
>
> Except in cases in which formal adoption by the Institute membership has been asked and secured, the authority of opinions reached by the committee rests upon their general acceptability.
>
> The committee contemplates that its opinions will have application only to items material and significant in the relative circumstances.[4]

Special notice should be given to the preceding paragraph dealing with the authority of committee opinions. Authority comes only through the willingness of the members of the AICPA to accept the opinions. In 1964, the Council of the AICPA adopted recommendations that, after 1965, all departures from APB *Opinions* and effective *Accounting Research Bulletins* should be disclosed in footnotes to the financial statements. Previously, all *Opinions* were considered to constitute the necessary substantial authoritative support for generally accepted accounting principles, but that other sources might exist. In the event of a practice contrary to the AICPA position, the member was to make a judgment as to the existence of some other authoritative support. If he was convinced that such authority existed, he was permitted to use the practice even though contrary to the AICPA.

Beginning in 1965, even if the accountant decided that a practice had substantial authoritative support outside the APB, he was required to disclose any departure from the official pronouncements if the effect on financial statements was material. The disclosure was expected to point out the difference between amounts shown on the financial statements and what would have been shown had the *Opinions* of the APB been followed. The failure to disclose such departure was considered to be substandard reporting and to be referred to the Practice Review Committee of the AICPA. In addition, as of January 1973, the failure to disclose a departure was made a violation of the Code of Professional Ethics.

The Securities and Exchange Commission

As discussed in Chapter 1, the Securities and Exchange Commission (SEC) has substantial legal authority. Created in 1933 as a result of the stock market crash of 1929, the SEC was given broad authority over the financial information which is to be provided to the public by those companies which plan to raise capital by a public sale of securities.

The Commission was given specific authority:

> . . . to prescribe the form or forms in which required information shall be set forth, the items or details to be shown in the balance sheet and earnings statement, and the methods to be followed in the preparation of accounts, in the appraisal or valuation of assets and liabilities, in the determination of depreciation and depletion, in the differentiation of recurring and non-recurring income, in the differentiation of investment and operating income, and in the preparation, where the Commission deems it necessary or desirable, of consolidated balance sheets or income accounts of any person directly or indirectly controlling or controlled by the issuer, or any person under direct or indirect common control with the issuer . . .[5]

Clearly, Congress has given the SEC substantial authority over financial reporting practices of those companies covered by the act. The

Commission, in the exercise of its authority relative to the development of accounting principles and practices, has followed its *Accounting Series Release 96*. In this release, the Commission explained that its role was "intended to support the development of accounting principles and methods of presentation by the [accounting] profession but to leave the Commission free to obtain the information and disclosure contemplated by the securities laws and conformance with accounting principles which have gained general acceptance."[6]

Thus the Commission has sufficient authority to enter actively into the development of accounting principles, but it has generally chosen to let the accounting profession take the lead. In recent years, however, the SEC has exhibited an increased interest in taking a more direct role in the development of accounting practices.

American Accounting Association

The American Accounting Association (AAA) is the successor of the American Association of University Instructors in Accounting. Because of the background of the association, the activities of the AAA are directed more toward academicians and theoretical problems rather than to practitioners and practical issues. Accordingly, the AAA published in 1936 *A Tentative Statement of Accounting Principles Underlying Corporate Financial Statements*. This was an attempt to set forth briefly (five pages) some of the bases upon which financial statements were prepared.

In this publication, the AAA made clear its view that this was a tentative statement, and that the most important applications of accounting principles lay in the preparation of published reports of profits and financial position. The AAA further stated that "Every corporate report should be based on accounting principles which are sufficiently uniform and well understood to justify the forming of opinions as to the condition and progress of the business enterprise."[7]

In 1941, the Executive Committee prepared a revision of the 1936 statement and dropped the "tentative" part of the title. Another revision appeared in 1948 with the title changed to *Accounting Concepts and Standards Underlying Corporate Financial Statements*. Between 1950 and 1954, a new Committee on Accounting Concepts and Standards prepared eight supplementary statements clarifying or expanding the 1948 statement. In 1957, another major revision was published, bearing the title *Accounting and Reporting Standards for Corporate Financial Statements*.

As is evident with each successive revision, the AAA committees became more positive in the titles chosen — as well as in content. Whereas the earlier statements were primarily concerned with specific rules and standards for corporate reporting, the purposes of the 1957 revision were "to present the concepts fundamental to accounting, and to suggest standards to which general-purpose reports to stockholders and others interested in corporate business enterprise should conform, and by which existing accounting practice may be judged."[8]

In 1966, a committee of the AAA issued a report titled, *A Statement of Basic Accounting Theory*. The committee recommended that four basic standards provide the criteria for evaluating potential accounting information. The standards suggested are relevance, verifiability, freedom from bias, and quantifiability. These standards and others will be discussed in the next section of this chapter.

Other Contributors

As discussed in Chapter 1, the Financial Executives Institute, the National Association of Accountants, and the Internal Revenue Service have also influenced financial reporting. In addition, several federal and state regulatory agencies have some impact on accounting. The most prominent ones are: the Interstate Commerce Commission, the Federal Trade Commission, the Federal Power Commission, and the Civil Aeronautics Board; however, the influence of these agencies has been indirect because they are concerned with the specific accounting problems related to their sphere of regulation.

CONTEMPORARY PROBLEMS IN THE APPLICATION OF GAAP

In this textbook, the manner in which financial information is accumulated and financial statements are prepared has been presented in the context of generally accepted accounting principles. Any textbook treatment of accounting applications must necessarily be limited to specific examples with few variables. Such examples serve to illustrate the various accounting principles and practices underlying financial statements; however, they cannot be viewed as totally representative of the real world applications of accounting principles.

In the real world, the large number of variables, which influence generally accepted accounting principles, gives rise to numerous problems and debate about their application. In some cases, there may even be a conflict between two generally accepted accounting principles, i.e., relevant data may not be totally objective. Thus financial statements are the result of numerous judgments as to the alternative accounting methods to be employed and the applicable generally accepted accounting principles.

In this section some of the current issues relative to generally accepted accounting principles will be presented.

Relevance

The basic purpose of financial accounting is to provide information about individual business enterprises that is useful in making economic decisions. The concept of usefulness is such an integral part of accounting that it is often implicitly assumed rather than stated as a principle.

The American Accounting Association uses the term *relevance* rather than usefulness. According to the AAA, "Relevance is the primary standard [of financial accounting] and requires that the information must bear upon or be usefully associated with actions it is designed to facilitate or results desired to be produced."[9]

The difficulty in the application of this principle lies in the determination of what constitutes relevant information. Since there are multiusers of the financial statements, what is relevant for one may be irrelevant for another. Furthermore, the principle of relevance may conflict with other principles, such as objectivity. For example, many people believe that current value information is more relevant than the historical cost information used in accounting; however, current value information, in many cases, cannot be objectively determined.

Historical Cost and Current Value

Accounting information is primarily historical. A major reason for this is the principle of objectivity. A general meaning of the term, as used in accounting, is that financial information is objective when it does not contain personal bias and can be substantiated by an independent investigator.

Broadly interpreted, this principle requires that only financial transactions between independent parties, which can be measured in dollars, are to be recorded in the accounting system. These transactions would normally be supported by such objective evidence as invoices for purchases, cancelled checks, and contracts.

There is a growing concern that historical cost is inadequate to properly present the financial position of the firm. This concern has, no doubt, been influenced by the rising inflation in recent years and its impact on the value of assets.

One proposed solution to this problem is the use of price level adjusted information, which was discussed in Chapter 9. The use of a price level index is the adjustment of the historical cost to common size dollars.

Therefore, this alternative is not a departure from a historical cost basis of accounting, but if the financial statements are presented in current dollars a step in the direction of current values may have been achieved. However, there are unresolved problems in using price levels pertaining to: what index to use; whether users would understand the adjustments; whether inflation has been severe enough to warrant adjustment; and whether adjusted information would be more relevant.

Many people believe that current values are more relevant than historical costs; however, there are substantial difficulties in attempting to use current value information for many assets. For example, what value would be placed on a 50% depreciated asset when there is no sale involving the asset. If a current value is derived then several questions may be raised:

1. How objective was the valuation process?
2. Is the valuation biased in any way?
3. Can the valuation be verified?
4. Is this information relevant for the user?

The answer to these questions depends upon the point of view of the respondent. There is a substantial amount of research currently underway directed towards, hopefully, answers to questions such as these.

Materiality

There is an almost unlimited amount of financial information which can be provided for any business; however, the reporting of every detail would be costly and could confuse users. Important facts, such as the purchase of a building, would be commingled with insignificant transactions such as the purchase of rubber bands.

In order to make financial statements more meaningful and to minimize cost, accountants should report only information which is material; however, there is no exact accounting definition of the term material. According to a committee of the American Accounting Association, "An item should be regarded as material if there is reason to believe that knowledge of it would influence the decision of an informed investor."[10]

Such a definition, obviously, places a burden upon the accountant to exercise judgment in the selection of material items. Certain items may be insignificant when viewed individually, but of major importance when combined with several other minor items. For example, if the combined total of the items exceeded 10% of net income, most accountants would consider it to be material. There is, however, no specific statement of the precise dollar amount or percentage that is to be considered material.

The problem in applying the concept of materiality arises from the fact that the various users of accounting information may be influenced differently by the same facts. Therefore, what is material to one user of the information may be immaterial to another. Thus the materiality concept is related to the same types of problems involved with relevance.

Materiality is also affected by factors other than dollar amount. For example, materiality may be involved in certain management plans, the nature of the item, current economic conditions, and other qualitative factors. Since materiality has many dimensions which are not fully understood, at present each situation must be judged individually.

Disclosure

Another problem area is deciding what constitutes adequate disclosure. The accounting profession has always maintained that all relevant information should be disclosed. Therefore, the concept of disclosure in financial reporting is directly related to the objectives of relevance and materiality. As discussed earlier, there are several problems inherent in these concepts; therefore, what constitutes adequate disclosure is also subject to interpretation.

Matching

The matching principle is directly related to the concepts of going concern and historical cost. Since accountants assume that the business will continue to operate in the future, liquidating values may not be relevant. In addition, in order for decision-makers to have relevant information, financial statements must be prepared on a periodic basis rather than waiting until the business ceases to operate.

The term *periodicity* refers to the segmentation of the firm's existence into time periods. Periodicity and its relationship to the going concern concept is illustrated in the following graph:

1 Year	1 Year	1 Year	1 Year

Life of business

The going concern principle is reflected by the continuum. The periodicity principle is represented by the one-year time intervals.

The problem created by periodicity is one of timing. For any accounting period, the accountant must determine the revenue items to be included in that period. Once this has been accomplished, the expenses involved in obtaining that revenue must be determined and matched with the revenue. Otherwise, the users of the financial statement would have misleading information because the expenses would not be related to the revenues; therefore, the profit reported would not reflect the results of operations for that period.

The recognition of revenue may occur at any one of four times. Those times, and the criteria, are:

1. At time of sale. This is the most common and is used when the price is known with certainty, when collectibility is reasonably assured, and when the related expenses can be estimated.
2. During production. This may be used when there is a firm price based on a contract or general business terms. For example, revenue is recognized periodically on long-term construction projects.
3. At completion of production. This may be used only when there is a determinable selling price or stable market price. For example, certain valuable minerals have fixed selling prices.
4. At time of cash collection. This may be used when it is impossible to estimate the collectible amount or when there will be additional expenses which cannot be estimated at the time of sale.

The matching principle is concerned with associating the cause and effect relationship of revenues and expenses. Therefore, if revenues are recognized in the current period, the expenses which were incurred in obtaining that revenue must be recognized. Certain expenses, such as sales commissions and the cost of goods sold, have a direct association with revenue. Other costs have an indirect association, and the accountant must allocate those costs in a systematic and rational manner to the periods benefitted.

Some examples of current problems involving matching include when to recognize revenue on land sales, long-term lease commitments, and whether expenditures for social improvements should be expensed or capitalized. Problems of this nature should serve as a reminder that the reported net income is the result of numerous decisions concerning the matching of revenues with expenses.

FUTURE PROSPECTS FOR THE RESOLUTION OF ACCOUNTING PROBLEMS

The accounting organizations have responded to the problems and criticisms of accounting in various ways. Perhaps the most significant profession-wide movement was the establishment of the Financial Accounting Standards Board. The FASB was established in 1973 and thus has had only a limited time to deal with accounting issues; however, a look at the organizational structure surrounding the FASB should provide evidence of the profession-wide involvement (see Figure 14-1).

The AICPA appoints the trustees of the Financial Accounting Foundation (FAF). The nine trustees are drawn from the following areas:

1. Chairman is the president of the AICPA.
2. Four members are CPA's in public practice.
3. One financial executive is nominated by the Financial Executive's Institute.
4. One financial executive is nominated by the National Association of Accountants.
5. One member is nominated by the Financial Analyst Federation.
6. One member is nominated by the American Accounting Association.

The FAF is charged with raising the capital necessary to operate the FASB and to appoint the members of the FASB. They also appoint and fund the Financial Accounting Standards Advisory Council.

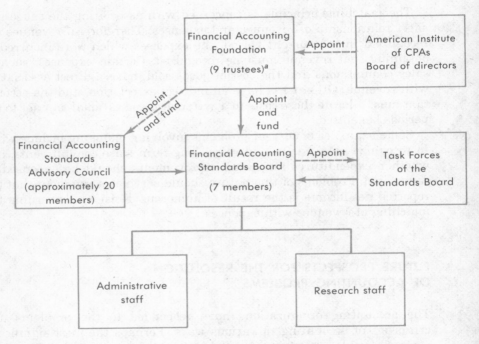

aFour of nine trustees are to be appointed from lists of nominees submitted, respectively, by the FEI, NAA, and AAA.

Source: Reports of the Study on Establishment of Accounting Principles

FIGURE 14-1

Financial Accounting Foundation organizational structure (Source: *Report of the Study on Establishment of Accounting Principles*)

The FASB is an independent body whose primary purpose is to issue statements of financial accounting standards. (Once appointed, the FASB is independent of the FAF, except for funding.) The members are drawn from:

1. Four members are CPA's.
2. The remaining three members are drawn from other areas related to accounting. These three individuals represent the viewpoints of education, government, and industry.

All of the other units are intended to support the activities of the FASB in the development of accounting standards. Only time will tell whether this profession-wide effort will produce satisfactory solutions to the various problems of financial reporting.

The preceding discussion should not be interpreted to mean that other organizations and individuals have discontinued their efforts to improve financial reporting practices. Rather there is evidence that the various organizations, e.g., SEC, AAA, etc., as well as individuals, are making greater efforts than ever to contribute to the improvement of the practice of accountancy.

SUMMARY

Generally accepted accounting principles, as is evident from the discussion, do not constitute a definitive set of rules for accounting. They only serve as guidelines to be used so that the information presented in the financial reports may be verified by someone else. Another accountant may not agree to the same dollar amount, but the differences should not be so great that the financial picture of the firm would be distorted.

While many organizations and individuals have contributed to the development of generally accepted accounting principles, the AICPA, AAA, and the SEC have been the most influential. At the present time, the Financial Accounting Standards Board has been entrusted with the primary responsibility for developing financial accounting standards. Hopefully this board will resolve many of the problems which now exist in the reporting of financial data.

FOOTNOTES

[1] American Institute of Certified Public Accountants, *Accounting Research and Terminology Bulletins* (Final Edition), 1961. *Terminology Bulletins*, paragraph 15.

[2] *Ibid.*, paragraphs 17 and 18.

[3] American Institute of Certified Public Accountants, Council of the Institute, Special Bulletin, *Disclosure of Departures from Opinions of the Accounting Principles Board*, 1964.

[4] American Institute of Certified Public Accountants, *Accounting Research and Terminology Bulletins* (Final Edition), 1961. *Research Bulletins*, pp. 8, 9, 10.

[5]*Securities Act of 1933.* U. S. Government Printing Office.

[6]U. S. Securities and Exchange Commission, *Accounting Series Release 96.*

[7]American Accounting Association Executive Committee, *Accounting and Reporting Standards for Corporate Financial Statements and Preceding Statements and Supplements,* 1957, p. 60.

[8]*Ibid.,* p. 1.

[9]American Accounting Association Committee to Prepare A Statement of Basic Accounting Theory, *A Statement of Basic Accounting Theory,* p. 7.

[10]*American Accounting Association Executive Committee, op. cit.,* p. 8.

QUESTIONS

14-1. What is the definition of *principle* which comes nearest to describing what most accountants mean by the word?

14-2. List the steps in the development of generally accepted accounting principles.

14-3. As applied to accounting practice, does the word principle connote a rule from which there can be no deviation?

14-4. What are the sources for determining whether an accounting practice has substantial authoritative support?

14-5. (a) What unit within the AICPA is considered to be the most influential and authoritative?

   (b) What determines the authority of the Financial Accounting Standards Board?

   (c) Discuss the treatment of departures from pronouncements by the APB and FASB.

14-6. (a) For what purpose was the Securities and Exchange Commission established?

   (b) Has the Securities and Exchange Commission entered actively into the development of accounting principles? Discuss.

14-7. (a) What is the American Accounting Association?

   (b) In 1966, the American Accounting Association issued *A Statement of Basic Accounting Theory* in which it recommended that four basic standards provide the criteria for evaluating potential accounting information. Name them.

14-8. Besides the major contributors, i.e., the American Institute of Certified Public Accountants, the American Accounting Association, and the Securities and Exchange Commission, there are other organizations and agencies which have contributed indirectly or minimally to the development of accounting principles. What are they?

14-9. According to the American Accounting Association, relevance is the primary standard of financial accounting. What is relevance as related to accounting and why is it so important?

14-10. In order to make financial statements more meaningful and to minimize cost, accountants should report only information which is material. Define material.

14-11. In order to match revenues and expenses for any particular accounting period, the accountant must decide what revenue items are properly includable in that period. List the times when revenue may be recognized and the criteria for each.

14-12. Define *objectivity* as it is used in accounting.

PROBLEMS

Procedural Problems

14-1. Tom Mason, owner of Mason's Department Store, a sole proprietorship, sells general merchandise. He hired Bruce Smith to maintain the accounting records. At the close of the first year of operations, Bruce prepared the following income statement and balance sheet:

Mason's Department Store
Income Statement
for the Year Ended December 31, 1974

Sales		$150,000
Less: Purchases		120,600
Gross profit		$ 29,400
Operating expenses:		
Salaries expense	$21,000	
Rent expense	6,000	
Advertising expense	1,200	
Property tax expense	800	
Miscellaneous expense	1,550	30,550
Net loss		$ 1,150

Mason's Department Store
Balance Sheet
December 31, 1974

Cash	$ 8,850
Equipment	9,000
Tom Mason, capital	$17,850

Because of the net loss shown on the income statement, Tom is not sure if he should continue in business. He has asked you to examine the statements and to make any necessary corrections. During the examination, the following facts are discovered:
1. Transactions were recorded only when cash was received or paid.
2. No closing entries have been made.

3. Tom started the business on January 2, 1974, by depositing $19,000 in a business checking account. No additional investment or withdrawal was made during the year.
4. Accounts receivable at December 31 amounted to $10,000.
5. Merchandise inventory at December 31 was estimated to be $22,000.
6. Store supplies of $400 were purchased during the year and were debited to Purchases. Supplies on hand at December 31 were $150.
7. Insurance premiums paid during the year amounted to $1,100 and were debited to Miscellaneous Expense. Premiums in the amount of $600 apply to 1975.
8. The equipment shown on the balance sheet was purchased for cash on January 2. Additional equipment of $6,000 was purchased on November 1 by issuing a 90-day, 6% note. No entry was made.
9. No depreciation has been recorded. The equipment is considered to have a ten-year life and no salvage value. The straight-line depreciation method is to be used.
10. Accounts payable at December 31 amounted to $1,800.
11. Property taxes owed but not paid on December 31 total $200.

REQUIRED
(a) Record the general journal entries necessary to correct the accounting records.
(b) Prepare a corrected income statement and balance sheet.
(c) On the basis of the revised statements, evaluate the operations of Mason's Department Store for the first year.

Conceptual Problems

14-2. • The Wildcat Insurance Company purchased letter openers for its officers. Office Equipment was debited for the $10 cost. Each year of the estimated ten-year life, Depreciation Expense is to be debited for $1. Wildcat's annual net income averages $500,000.
• The Enormous Steel Corporation debits an expense whenever it buys automobiles for the use of its salesmen even though company policy is to use automobiles for three years. The company buys ten autos per year at a cost of approximately $3,000 each. The annual net income averages $50,000,000.

REQUIRED. Discuss the preceding practices in terms of the generally accepted accounting principle of materiality. Is each of these practices acceptable?

14-3. The Better Lettuce Company had an excellent year in 1973. In order to report a lower profit, the books were not closed on December 31. All checks written during the first two weeks of January, 1974 were recorded as 1973 expenses.

REQUIRED. What generally accepted accounting principle(s) has been violated? Explain.

14-4. On January 1, 1974, Jones Manufacturing Company purchased a machine for $40,000. On December 31, 1974, the machine's fair market value had increased to $50,000. Since there was no decrease in value, Jones recorded no depreciation. Instead, the equipment account was increased by $10,000 and a gain was recognized.

REQUIRED. Does this practice adhere to generally accepted accounting principles? If not, explain which principle(s) has been ignored.

14-5. When Cooling Systems, Inc. has a bad year, its bookkeeper tries to make things appear better by not recording depreciation. In good years, he records extra depreciation in order to bring the accumulated depreciation account up-to-date.

REQUIRED. Is such a practice acceptable considering the generally accepted accounting principle of consistency?

14-6. The chief accountant for the Specific Motors Corporation has presented the following plan to the company's board of directors. Each year the company produces automobiles which are shipped to dealers across the country for ultimate sale to the public. The revenue recognition procedure utilized in the past was to recognize revenue at the time an auto was shipped to a dealer. The chief accountant has been with the company for five years and has observed that every auto manufactured has been sold to a dealer. He thus suggests that the company begin to recognize revenue at the point when production is completed. He reasons that Specific Motors Corporation has never had to discard any of the autos they manufacture due to overproduction or lack of demand.

REQUIRED. State whether you agree or disagree with the new proposal suggested by the chief accountant. Support your position.

14-7. As a new employee of the Hi-Style Fashion Company, Susan Mac was asked to review the accounting procedures utilized by the company. Ms. Mac is a recent accounting graduate and her assigned task was twofold. She was asked to make sure the company used the accounting procedures which reflected the best picture of the company's operation; however, all procedures used were to follow generally accepted accounting principles. In accordance with these constraints, Ms. Mac developed the following set of policies for the accounting department to follow:

- When the price of inventory items was rising the first-in, first-out method was to be used. When prices were on the decline the last-in, first-out method was to be used. Profits would always be maximized if this policy was followed.

- Fixed assets were to be depreciated using the double-declining balance method for the first half of the asset's useful life. During the remainder of the useful life, the straight-line method was to be used with careful attention paid to the asset's depreciable cost so that total depreciation charges never exceed this amount.

- The estimated bad debt expense should be recorded at twice the anticipated amount each year to be sure to provide for any unexpected bad debts. At the end of each year the excess provision should be credited to an account entitled, Gain from Bad Debts Not Realized. This policy is in accordance with the accounting principle of conservatism which states that accountants should provide for all losses and anticipate no gains.

- The accounts Prepaid Insurance and Prepaid Rent should be discontinued. Any time insurance and rent are paid in advance the entire amount will be debited to expense. This policy is in accordance with the going concern concept since it is assumed that the company will continue in operation indefinitely; thus these payments will definitely become expenses.

REQUIRED. Comment on the policies Ms. Mac has instituted based upon the constraints she was instructed to operate within.

14-8. You are employed as an accountant for a medium-sized manufacturing firm. Recently the president called you into his office and asked you to comment on a conversation he had with his stockbroker. The stockbroker, in helping the president analyze the financial statements of other manufacturing firms, mentioned that financial statements are composed of such a great number of estimates that the true statement totals may vary as much as 50% in either direction. The president is quite shocked by the extensive amount of guess work which the broker claims is a large part of financial statement preparation.

REQUIRED. State the major areas of financial reporting wherein estimates are required. Indicate the reasons for such estimates and the problems that would result if estimates were not made.

15

MANAGERIAL ACCOUNTING: AN OVERVIEW

EDUCATIONAL OBJECTIVES

The material in this chapter is designed to achieve several educational objectives. These include:

1. A knowledge of how managerial accounting differs from financial accounting.

2. A basic understanding of budget activities, such as profit planning, capital budgeting, and flexible budgeting.

3. An insight into the various mathematical techniques that may be used in accounting.

4. A knowledge of how internal auditing activities differ from the activities of the external certified public accountant as well as the relationship between these two audit activities.

5. A basic knowledge of how standard costs are developed and applied.

Managerial accounting is distinguished from financial accounting primarily on the basis of the users. Financial accounting, as discussed throughout the book, is concerned with providing reliable quantitative information to a multitude of users, most of whom are external to the firm. These external users include investors, creditors, governmental units, and employees through their representatives, the labor unions. Since these users are external to the firm, they generally cannot directly obtain the information they desire; therefore, the external financial statements are prepared according to generally accepted accounting principles. These statements are then audited by Certified Public Accountants, who express their opinion as to the fairness of the statements. Thus the external users are provided assurance, through an independent agent, that the financial statements are reliable.

These external users obviously must make decisions relative to their particular interest, e.g., to invest or disinvest, whether or not to grant credit, etc. Such decisions, while extremely important, may be made infrequently. The point is that external users may review the financial statements, make a decision not to invest, and that will be the end of that individual's use of the financial statements of the given firm.

Within the firm, however, management is constantly making decisions relative to raising capital, acquiring operating facilities, production or acquisition of inventory, and the sale of inventory. These numerous activities must be planned and they must be controlled. Managerial accounting is concerned with all of the financial reporting activities necessary for managerial decision-making.

The purpose of this chapter is to provide an overview of managerial accounting as a prelude to a complete course on the subject. In many schools, the second course in accounting is concerned with the managerial aspects. The topics to be discussed in this chapter are not intended to be all-inclusive; rather, they are representative of the activities involved in managerial accounting. The three major aspects of managerial accounting to be presented are:

1. Planning
2. Technical considerations
3. Control

PLANNING

The planning function of management involves not only the day-to-day scheduling of events but also the long-range plans that affect the objectives of the firm. Many variables are involved in making these decisions, and those to be discussed are profit planning, forecasting, capital budgeting, behavior of costs, and the effects on the human side of the business.

Profit Planning

A well-managed firm does not haphazardly achieve a certain profit level. The actual earnings per share are the result of plans made many months prior to the final determination of EPS. The actual EPS may not be exactly the expected profit, but they should be in the range originally anticipated. The reason for the variance is obviously the difficulty inherent in attempting to anticipate all possible contingencies that may affect profits. Thus the original planned profit is generally a range based on assumptions concerning the various aspects affecting a particular firm.

Profit planning does not operate independently of other considerations within the firm. Profits cannot be obtained without capital resources, and these resources cannot be maintained without sufficient profit. As an individual, you are faced with the same type of problem. You make commitments to purchase items and to pay for them in the future, e.g., house, car, etc. However, you cannot normally purchase these items without some income and yet these items may be an integral part of your effort to earn the income to pay for these items. Thus profit planning must be a reflection of the profit level that can be achieved by the firm.

Once the profit range has been determined, management attempts to organize all activities of the firm to achieve this goal. Thus, as a broad term, profit planning includes both the planning and coordination of all operations towards a profit objective and the evaluation and planning of individual projects in terms of their contributions to this objective.

Forecasting

Inherent in the planning process is the forecast of events occurring in the future and their impact on the firm. These forecasts normally include projections of:

1. The economy, both short-run and long-run, and its impact on the industry and the firm.
2. Market analysis of the type and volume of sales for the industry and the firm.
3. The nature of competition, both within and outside the industry.
4. Production capabilities.
5. Labor and material availability.

These forecasts should be as specific as possible because if the forecasts are specific, the projections can form the basis for budgets. These budgets then form the basis for organizational goals to be achieved, e.g., production efficiency, labor utilization, and sales. Thus the planning function becomes a control mechanism if the plans are specific; however, to properly develop budgets, one must analyze the costs to be incurred since the nature and behavior of these costs will influence the magnitude of the operations and present constraints for the budgets.

Capital Budgeting

Profit planning, as indicated, cannot operate independently of capital resources. As a major ingredient in the planning process, the purchase of capital resources, or capital budgeting, will now be discussed.

A capital budget is a plan to purchase long-term assets. Capital budgeting refers to the decision-making process necessary to allocate the firm's scarce resources among various alternatives. Capital budgeting is important to management because:

1. Large sums of money are involved in purchasing capital assets. Consequently, mistakes can have serious effects on the efficient operation of the firm.
2. Since the effects of the purchases extend over a long period of time, the existence of a firm may be in jeopardy if an incorrect forecast of asset needs is used.

In addition to the purchase of fixed assets, capital budgeting techniques are also used for the replacement of assets, determining the efficient size of a plant, lease or buy decisions, and refunding of debt. In capital budgeting, cash flow, rather than net income, is a significant factor in the decision to accept an alternative. Cash flow is used because cash is necessary for the payment of dividends and for the payment of the assets purchased.

There are four items that must be determined for capital budgeting decisions:

1. Amount of cash benefits
2. Timing of cash benefits
3. Risk involved in the attainment of the benefits
4. Cost of the asset

The majority of capital budgeting techniques involve relating, in some manner, the benefits to be received with the cost of the asset.

The four techniques of capital budgeting normally used are:

1. *Payback.* In this technique the cash flow from the investment is related to the cost in order to determine the time involved before the cash inflow equals the cost of the investment. For example, if the cash inflow was $2,000 a year for five years and the cost of the

asset $6,000, the payback period would be three years. This method is widely used as a simple measure of the return on the investment.

2. *Accounting rate of return.* This technique relates the average net income from the asset to the average cost. This method does not consider cash flow, time value of money, nor any risk factor. For these reasons, this method must be used with extreme caution.

3. *Present value.* This technique relates the cash flow to the cost by multiplying the flow for each year by an applicable present value factor. The present value factor to be used is the minimum rate acceptable for an investment of this type. If the present value of the inflows is greater than the cost, the investment is acceptable because the rate of return on the asset is greater than the minimum required.

4. *Discounted cash flow (DCF).* This technique, like the present value method, relates the cash flow to the cost; however, in this method, the actual rate of return is calculated rather than determining if the rate is greater than an acceptable minimum level.

Further discussion of these techniques must be reserved for a later course; however, this discussion should indicate that, as with generally accepted accounting principles, there are different methods available for the accountant to assist management in making decisions.

Cost Behavior

Planning decisions require that attention be given to the behavior of costs. The expression *cost behavior* refers to the fact that certain costs tend to change in relation to the quantity of activity while other costs remain relatively constant. An understanding of such cost behavior is necessary for decision-making, and information about cost behavior can be obtained from the accounting records.

Fixed Costs. Those costs which remain relatively constant when the activity level changes are referred to as *fixed costs*. Among these costs are such things as property taxes, insurance on the plant and equipment, depreciation on a straight-line basis, and the salary of management personnel. These costs remain constant within a certain range of activity. If the level of activity is increased beyond the upper limit then, clearly, the plant will need to be expanded and there will be a new amount of fixed cost.

Variable Costs. When costs tend to change in direct proportion to the change in activity level, they are said to be *variable costs*. Perhaps the most common variable costs are the materials which go into the manufacture of a product and the labor directly associated with the productive process. As the level of activity increases, the quantity of material and labor necessary will tend to increase at the same rate.

A major problem for managerial accountants arises in connection with cost behavior because not all costs are clearly fixed or variable. Certain costs have elements of both and may be referred to as *semivariable* or *semifixed*. In any event, for decision purposes, such costs must be separated into the fixed and variable portions. There are certain techniques, e.g., scattergraph, least squares, etc., which can be used to assist the accountant in making these allocations.

The separation of costs into fixed and variable aids management in attempting to predict what will happen to costs if certain activity levels are achieved. Without such information, management could only guess at what would happen to profits if certain changes relative to production were implemented.

Behavioral Considerations

In the not too distant past, economic and organization theory asserted that the principal objective of business was profit maximization. According to this theory, employees were motivated to perform primarily by money and they had little interest in the achievement of the firm's goals. The major role of management, in this framework, was to obtain the maximum profit, and the role of managerial accounting was to assist management in achieving the firm's goal.

The findings of behavioral scientists have, in recent years, begun to have an effect on organization theory and, consequently, management and managerial accounting. According to the newer theory of organizations, the firm is made up of individuals; therefore, the goals of the firm must necessarily reflect the goals of individuals. In this theory profit is important, but it is not the only goal and there is an effort to obtain a satisfactory, rather than maximum, profit. There are other important goals to be achieved, e.g., the psychological and social needs of the participants.

This newer theory places additional burdens upon management because of the need to find a balance between profits and these other goals. As a result, managerial accountants should be more versatile than in the past. They should recognize the needs of individuals, and those needs include the desire to participate in those decisions which concern them personally. For example, under the older theory, the accountant and management would prepare budgets which were intended to produce maximum profits. The individuals responsible for each area of activity were given their budget and they were expected to achieve the specified goals. Under the newer theory, those individuals are consulted and provide input for the development of the budget. Accountants, therefore, should know how to communicate with all levels and segments within the firm.

One other illustration of behavioral considerations is human resource accounting. This is emerging from the growing realization that a firm's employees are often its most valuable assets, yet they are not shown on the balance sheet. Assets have been defined as future service potential and clearly the hiring, training, and developmental costs of employees were undertaken with the expectation of future benefits. Some firms are beginning to treat such expenditures as assets which are to be amortized over the expected employment period.

TECHNICAL CONSIDERATIONS

Now that the planning aspect of managerial accounting has been discussed, attention will be directed to certain techniques employed within the organization. These techniques include production costing, under both job-order and process, and certain mathematical and computer applications.

Production Costing

A major difference between accounting for a manufacturing and a retail firm pertains to inventory. In a retail firm, the merchandise is purchased for resale and the cost includes the invoice price less any discounts, plus other acquisition charges, such as freight. Thus the only costing problem is in connection with the assumption as to the flow of cost, i.e., LIFO, FIFO, weighted average, etc.

In a manufacturing firm, the cost of inventory must be determined on the basis of the various inputs to the productive process as well as the assumption as to flow of cost. The inputs for which the accountants must have records are commonly grouped under the headings: material, labor, and overhead. The first two items were explained previously as elements of the end product or items directly associated with the productive process. *Overhead* is the term applied to all costs which are necessary for, but only indirectly associated with, production, e.g., depreciation, property taxes, utilities, etc.

When raw materials are purchased in a manufacturing firm, the entry is a debit to Materials and a credit to Cash. This is basically the same entry that a retailing firm makes for the purchase of inventory:

Materials   XX
    Cash   XX

However, when salaries are paid in a manufacturing firm, the entry is a debit to Direct Labor for the salary paid to production employees and a debit to Overhead for the salary of management employees who are only indirectly involved in production.

Direct Labor	XX	
Overhead	XX	
Cash		XX

The salary cost is not considered an expense at this point. In like manner, when taxes, utilities, etc. are paid they are recorded as Overhead and not as expenses.

As the productive process begins, the expenditures which have been accumulated in the Material, Labor, and Overhead accounts are transferred to a Work-In-Process inventory account. When the manufacturing process is completed, the costs associated with the finished product are transferred from the Work-In-Process account to the Finished Goods inventory account. The sequence of these cost transfers is depicted in Figure 15-1.

The accumulation and transfer of costs may be accomplished through either job-order or process costing techniques. The selection of the approach to be used depends primarily upon the nature of the production process.

Job-Order Costing. When each unit or batch of output is different, the costs of material, labor, and overhead must be accumulated by each job in order to determine the cost of producing the particular unit or batch. This job-order approach to costing is commonly used in such industries as job printing, construction, and specialty furniture manufacturing.

The record keeping in a job-order system is a major task if the firm has many jobs. There must be a record of all material and labor devoted to the particular job, and a portion of the total overhead must be assigned to each job.

Process Costing. In many manufacturing activities the units of output are the same; therefore, the costs may be accumulated by processes. Process costing systems are commonly found in mass-production industries such as textiles, soap, cigarettes, and breweries. In a process cost system, material, labor, and overhead costs are assigned to each process in the manufacturing cycle. The costs assigned to a given process are then divided equally among the units going through that process since all units require the same effort.

Mathematical Considerations

In recent years, mathematics has been used to solve complex problems found in business; however, the advocacy of the use of mathematics may

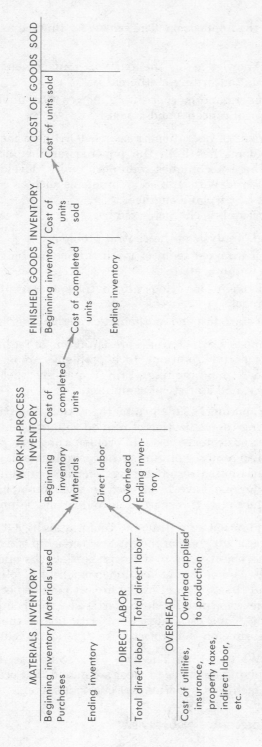

FIGURE 15-1

be greater than the application. The reason for this seems to result from two factors:

1. Lack of knowledge on the part of management and internal accountants concerning mathematics.
2. Lack of understanding concerning the practical applications of the mathematical concepts and models.

The development of operations research (OR) as a means of solving managerial problems has been the impetus for the increased use of mathematics. These techniques were originally developed by Great Britain during World War II to solve problems such as convoy size and repair schedules for airplane engines.

Operations research is characterized by:

1. A thorough analysis of the central problem.
2. The use of a mixed team of mathematicians, biologists, sociologists, accountants, etc.
3. The development of various alternatives as a result of the mixed team approach.
4. The use of statistics and mathematics as the basis for the analysis.

OR develops models, which are representations of reality, to evaluate the various alternative solutions to a problem. Among the common models are linear programming, PERT, queuing theory, statistical probability theory, and statistical sampling.

Linear Programming. This is a mathematical approach to business problems involving the utilization of limited resources in such a way as to increase profits or decrease costs. In problems of this type, an objective function (maximize profit or minimize cost) is to be maximized subject to the constraints of limited resources, e.g., machine capabilities. Applications of this technique have been used in such problems as machine scheduling, product mix, transportation routes, and shipping schedules.

Program Evaluation and Review Techniques (PERT). This is a formal, probabilistic diagram of the interrelationship of a complex series of activities. The objective is to discover bottlenecks and determine if certain activities can be completed independently of others. If this is possible, the total time involved in the project may be decreased.

This technique was initially developed and used in missile research and development. Time estimates were critical and the use of PERT enabled the total time on the project to be drastically reduced.

Queuing or Waiting-Line Theory. This technique deals with the problems of supplying facilities to meet a demand that occurs in uneven spurts, e.g., toll facilities, supermarket checkouts, etc.

Statistical Probability Theory.   This is a rigorous approach to the quantification of uncertainties. Since the manager is seldom certain concerning future events, he can make choices using expected values of various outcomes.

If the possibility of earning $20,000 is 50%, the expected value is $20,000 \times 0.50 = \$10,000$. This means that given the laws of chance, the expected value of receiving the $20,000 is, on the average, $10,000. If an alternative existed for earning $15,000 with total certainty, the rational decision would be to choose the $15,000.

The difficulty in this technique, naturally, is the assignment of probabilities to all the possible events under consideration. These probabilities should be based on historical evidence as modified by business judgment; however, if no objective information exists, subjective estimates may be used based on management's best judgment. The use of these subjective probabilities forces the executive to make explicit his intuitive feelings.

Statistical Sampling Procedures.   A large portion of the accountant's time is spent preparing reports from samples. As a consequence of this, the use of statistical sampling has increased in business. Sampling procedures can be utilized in a wide variety of situations such as:

1. Aging accounts receivable (see Chapter 6).
2. Estimating physical inventories (see Chapter 7).
3. Test checking for clerical errors.
4. Assisting the internal auditor in determining adherence to the system developed for the particular area.

There are other models such as simulation, dynamic programming, regression analysis, EOQ models, etc. that could be discussed. Hopefully, the discussion presented indicates the wide range of possible uses of mathematics to assist management in achieving the goals of the firm.

Computer Considerations

Computers have contributed to the development of OR techniques, assisted accountants in providing information for management, and tended to centralize the data base. Computers receive raw information as input, perform arithmetical computations, make logical decisions with respect to these data, and issue the processed information as output. This is accomplished with electronic speed usually limited only by the mechanical abilities of the input and output devices. This speed has made possible the evaluation of many more alternatives and freed middle-management from most data processing concerns. Optimal decisions are made easier because more alternatives are evaluated.

The total use of the computer is only for a few large firms. The majority of firms do not have the need for, or capabilities of, fully utilizing the computer; however, for the average firm there are still many alternatives available for a profitable use of the computer. These areas include inventory control, accounts receivable processing, fixed asset control, and payroll accounting.

An example of the activities of a computer for a payroll system will be illustrated by a flow chart (see Figure 15-2). The payroll process could take a great deal of time, but through the use of the computer both time and the number of clerical errors are minimized.

CONTROL

All of the best made plans of management may prove fruitless unless control measures are employed. In a managerial accounting context, control means having procedures for detecting when plans are not being followed and a reporting system to convey information to the responsible person. The aspects of control to be discussed in this section are: flexible budgets, standard costs, internal auditing, and responsibility accounting.

Flexible Budgeting

The concept of budget preparation was mentioned in the context of the behavioral consideration in accounting; however, the actual procedures involved were not presented. A budget is a *plan*; therefore, when a budget is prepared all factors which will influence the outcome must be considered. A budget may be prepared for each of many small units, for a division, or for the entire company. Normally, budgets are built from the smaller units to the total plan for the firm.

In order to illustrate the concept of flexible budgeting, assume a production unit within a manufacturing firm. Some of the factors which will be considered include: the price of raw material, labor pay rate, and the overhead rate which will be allocated to the units of output. Once the rates have been determined, the expected activity level must be considered.

A budget is prepared on the basis of the predetermined rates multiplied by the expected activity level. The raw material and direct labor costs vary with the level of output; therefore, if the level changes so will the costs, and the budget will be inappropriate. On the other hand, overhead costs have elements of fixed costs; therefore, the overhead rate is based upon a certain level of output such that the rate used will apply all of the overhead. If the level of production changes from the expected, the rate will either be too high or too low for application of the fixed overhead costs.

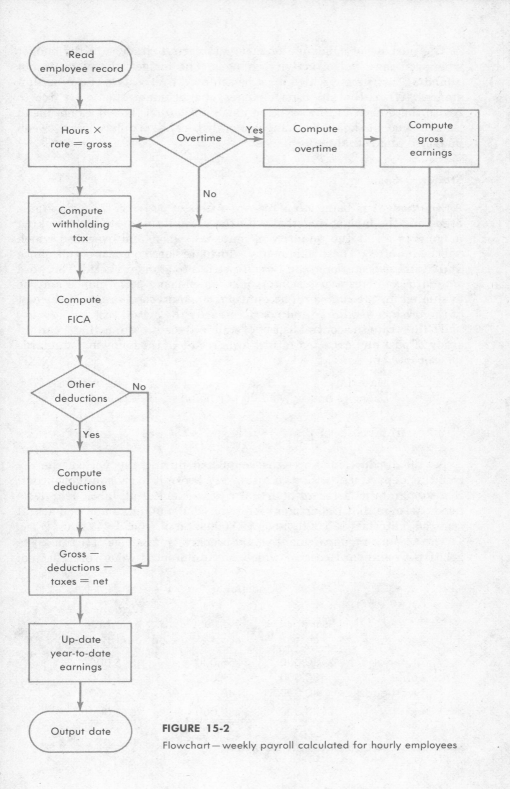

FIGURE 15-2

Flowchart—weekly payroll calculated for hourly employees

The purpose of a flexible budget is to permit changes in the budget whenever the level of activity changes. The budget, therefore, is not a binding instrument which must be achieved irrespective of the circumstances. When the situation changes, management should be free to revise the budget in line with the new information. This does not mean that the budget should be changed merely because of differences between budgeted and actual performance.

Standard Costs

As the budget is being prepared for a unit of activity, the individuals preparing the budget, together with the accountant, are clearly making judgments about the quantity of material, labor, and overhead which will be required. These judgments, which are based on past performance and future expectations, can be converted into standard costs. The word *standard* connotes a criterion against which some performance may be evaluated. In the context of accounting, standard costs are what the cost of the product *should be* and not necessarily the actual cost.

To illustrate the control aspect of standard costs, assume that a careful study of past and expected future costs result in the following standards for Department A:

Materials	$ 5 per unit
Labor (2 hours @ $4.00 per hour)	$ 8 per unit
Overhead	$ 4 per unit
Total cost per unit	$17

As the productive process is completed during the period, the accounting department assigns a cost of $17 per unit to each unit of output; it also maintains a record of actual costs for material, labor, and overhead. Assume that Department A produced 10,000 units and had actual costs of: materials $55,000; labor $79,000; and overhead $48,000.

The accounting department would prepare an analysis of actual costs relative to standard cost in which any differences would be reflected:

Department A

	Standard cost	Actual cost	Variance favorable and (unfavorable)
Material	$ 50,000	$ 55,000	($ 5,000)
Labor	80,000	79,000	1,000
Overhead	40,000	48,000	(8,000)
	$170,000	$182,000	($12,000)

The manager of Department A is expected to explain why the variances occurred. He should be assisted in his explanation by the accountant and reference to the detail records of actual costs.

Internal Auditing

A control device used in virtually every large corporation is the internal audit staff. Internal control is also important for smaller firms, but such firms may need only one or two people involved in internal auditing, whereas the larger firms may have a larger internal audit staff.

Internal auditing is defined by the Institute of Internal Auditors as "the independent appraisal activity within an organization for the review of accounting, financial and other operations as a basis for service to management. It is a managerial control, which functions by measuring and evaluating the effectiveness of other controls."

The Statement of Responsibilities of the Internal Auditor of the Institute of Internal Auditors describes the internal auditing activities as:

1. Reviewing and appraising the soundness, adequacy, and application of accounting, financial, and operating controls.
2. Ascertaining the extent of compliance with established policies, plans, and procedures.
3. Ascertaining the extent to which company assets are accounted for, and safeguarded from losses of all kinds.
4. Ascertaining the reliability of accounting and other data developed within the organization.
5. Appraising the quality of performance in carrying out assigned responsibilities.

The list of duties and responsibilities includes activities similar to those performed by the external auditor, the certified public accountant. The internal auditor is concerned more with the internal control aspects while the CPA is primarily concerned with the fairness of the financial statements; however, fairness can only be judged after the adequacy of the internal controls has been determined. The internal auditor is, therefore, a major contributor to the efficient operation of the firm and the effort of the external auditor.

Responsibility Accounting

Another control mechanism, which has been employed in many modern firms, centers around the areas of responsibility that exist in an organization. Responsibility accounting represents the accountant's attempt to classify the results achieved by an organization according to decision centers. These responsibility centers can be distinguished by departments, divisions, territories, or individuals. The specific centers used within a firm are a function of the needs of the individual organization.

Responsibility accounting can be viewed both as a system of cost and profit centers. In a *cost center system*, the only responsibility is the control of costs. This situation is applicable to those departments, etc. that do not generate revenues, e.g., maintenance. In a *profit center system*, both the control of costs and the generation of revenues are the responsibility of the center. This situation is applicable to a selling department.

Responsibility accounting has affected business in two ways. First, management must take a close look at costs and revenues and determine the individual or individuals responsible rather than making arbitrary allocations. Second, the assignment of costs to the individual managers who have control over their incurrence is a factor in encouraging these managers to motivate their subordinates. Managerial performance in this regard is measured by the accounting reports, which are likely to be an incentive for the effective motivation of the managers.

Thus responsibility accounting has assisted in moving budget preparation to the lowest possible level. In addition, this has given the workers some feeling of control over their activities and how they are to be evaluated in the performance of their work.

SUMMARY

The purpose of this chapter was to provide an overview and brief introduction to the activities of managerial accounting. The presentation centered around the accountant's assistance to management in the areas of planning and control. Certain techniques of managerial accounting, such as job-order and process costing, were discussed together with some suggestions as to mathematical and computer applications. There are clearly many other activities relating to managerial accounting since the role is basically one of serving the needs of management for quantitative information. A more complete discussion of the topics in this chapter, as well as other activities of managerial accounting, may be found in managerial accounting texts.

QUESTIONS

15-1. How is managerial accounting distinguished from financial accounting?

15-2. Why is profit planning important?

15-3. Forecasting involves projection of future events. What type events are included in forecasts?

15-4. Why is capital budgeting important to management?

15-5. List the items that must be determined for capital budgeting decisions.

15-6. What are the techniques for capital budgeting? Describe each.

15-7. Explain the difference between fixed and variable costs.

15-8. Describe the difference between traditional theory of management and the behavioral theory of management.

15-9. What has been the effect of the behavioral theory of management on managerial accounting?

15-10. Explain the difference in accounting for inventory in a manufacturing and a retailing firm.

15-11. What is the entry made for placing $1,000 of raw materials, $2,500 of direct labor, and $2,000 of overhead into Work-In-Process?

15-12. Discuss the difference between job-order and process costing.

15-13. What are the characteristics of operations research?

15-14. What is the expected value of earning $2,000 with a probability of 60%?

15-15. How does a flexible budget differ from a static or rigid budget?

15-16. What is the meaning of standard as used in managerial accounting?

15-17. What is meant by responsibility accounting?

PROBLEMS

Procedural Problems

15-1. Machine A costs $30,000 and will produce a cash inflow of $9,000 for 8 years. Machine B costs $40,000 and will produce a cash inflow of $10,000 for 12 years. If there is enough cash to buy only one asset, which asset should be purchased, and why?

15-2. The Wagner Corporation uses a standard cost system. The following data are available for the milling department for the month of August:

Current Standards

Material	$6 per unit
Labor	3 hours @ $3 per hour
Overhead	$5 per unit

The production for the month was 20,000 units and the actual costs were $117,000 material, $186,000 labor, and $110,000 overhead.

REQUIRED. Prepare a schedule showing the variances from the standard costs.

Conceptual Problems

15-3. As the head of the accounting department, you have recommended to the president that an independent internal audit department be established. The president remarks that the accounting function should be responsible for developing procedures to prevent fraud and embezzlement; therefore, he sees no need for a new department. Prepare a response to the president in order to convince him of the usefulness of an independent internal audit staff.

15-4. A major problem in many organizations is the lack of involvement in budget preparation by middle-management and a feeling of frustration concerning performance reports. In an effort to counter this problem, many firms have adopted the concept of responsibility accounting. Describe the basic characteristics of responsibility accounting and explain why such a system may overcome managerial frustrations.

15-5. The Merril Apparel Store has the following information concerning the operations of their three departments last year:

	Men's clothing	Women's clothing	Children's clothing
Sales	$185,000	$234,000	$96,000
Cost of sales	91,000	116,400	52,500
Gross profit	$ 94,000	$117,600	$43,500
Expenses:			
Selling:			
Advertising	$ 1,500	$ 1,500	$ 1,000
Delivery	5,000	6,000	4,000
Rent	15,000	19,000	11,000
Sales salaries	9,000	8,000	6,000
Total	$30,500	$34,500	$22,000
Administrative:			
Insurance	$ 8,000	$ 8,000	$ 7,000
Depreciation	6,000	6,000	4,000
Utilities	5,000	5,000	5,000
Clerical salaries	12,000	12,000	10,000
Total	$31,000	$31,000	$26,000
Total expenses	61,500	65,500	48,000
Income before taxes	$ 32,500	$ 52,100	$ (4,500)
Income taxes (50%)	16,250	26,050	2,250[a]
Net income (loss)	$ 16,250	$ 26,050	$ (2,250)

[a]Tax credit due to loss

The president of the company suggests that the children's department be discontinued as it has shown a similar loss for the past three years. You, as chief accountant, have been asked to prepare a report showing the profits that would have resulted had the children's department been discontinued. Some additional information concerning the expenses of the children's department is as follows:

- Two sales clerks are employed in the department on a part-time basis. Only one would be retained at the same wage to sell in the women's department.
- Advertising and delivery expenses would be unnecessary. The area occupied by the children's department would be split equally between the other two departments.
- Only half the insurance payment would be required, clerical salaries of the children's department would be reduced by 60%, the depreciable fixtures would be divided equally between the other two departments, and utilities cost would remain the same.
- Income from the two remaining departments should increase by $26,000 after taxes if the department is discontinued.

REQUIRED. Prepare a report in accordance with the president's request including your recommendation.

INDEX